AF360741

Genetic Diversity of Soil Bacterial Communities

Genetic Diversity of Soil Bacterial Communities

Editor

Carmine Crecchio

MDPI • Basel • Beijing • Wuhan • Barcelona • Belgrade • Manchester • Tokyo • Cluj • Tianjin

Editor
Carmine Crecchio
Università degli Studi di Bari Aldo Moro
Italy

Editorial Office
MDPI
St. Alban-Anlage 66
4052 Basel, Switzerland

This is a reprint of articles from the Special Issue published online in the open access journal *Diversity* (ISSN 1424-2818) (available at: https://www.mdpi.com/journal/diversity/special_issues/soil_bacterial).

For citation purposes, cite each article independently as indicated on the article page online and as indicated below:

LastName, A.A.; LastName, B.B.; LastName, C.C. Article Title. *Journal Name* **Year**, *Volume Number*, Page Range.

ISBN 978-3-03943-743-6 (Hbk)
ISBN 978-3-03943-744-3 (PDF)

Contents

About the Editor

Carmine Crecchio obtained his Ph.D. in Molecular and Cellular Biology in 1991. He has been an associate professor in Agricultural Chemistry at the University of Bari Aldo Moro since 1998. His main recent research activities deal with: (i) the investigation of biochemical, microbiological and molecular paramenters that affect soil quality and fertility as a consequence of different anthropogenic activities and agricultural soil managements, focusing in detail on the composition and functioning of bacteria communities by metagenomic approaches; (ii) the isolation and characterization of plant growth promoting bacteria and their use in agriculture as biofertilizers. He is the author of around 70 research papers and reviews, published in international journals; he is a reviewer for the most relevant journals in the field, member of the Editorial Board of Soil Biology and Biochemistry and evaluator of scientific projects and grant proposals for many international organizations.

Preface to "Genetic Diversity of Soil Bacterial Communities"

Soil is an important natural resource and has a key role in the biosphere, as most of the carbon and nutrient fluxes occur in the top 10 cm of the soil profile; it is a species-rich habitat that provides support for plant growth and health and consequently affects human activities. Although broadly homogeneous in the landscape, soil is extremely heterogeneous on a microbial scale. In fact, soil supports taxonomic and physiologic microbial diversity, which is regarded as more extensive than that of any other group of organisms and considered vitally important to the maintenance and sustainability of the biosphere.

Soil microbiome has a main role as a driver of the living soil that, in turn, contributes to key life support functions. Anthropogenic activities using soil, such as intensive agriculture, pose a burden on soil functioning. Therefore, it is important to find out good indicators able to detect deleterious changes and thus soil quality.

Looking within the "black box", as soil has been regarded in the last decades, overcoming its inaccessibility, and understanding its microbial composition and functioning, are challenges for scientists. In particular, if it is important to investigate the genetic diversity of microbial populations, it is also fundamental to understand the link between the major functions of microbial biomass and its species composition.

This book investigates the roles that various grass species and their functional forms play in modifying soil bacteriobiome and enzymatic activity, how plant genotypes can influence the bacterial communities related to phosphate mineralization and nitrogen fixation in the rhizosphere, the effects of crops and their cultivation regimes on changes in the soil microbiome and of three different long-term land use intensities on soil biochemical, microbial, and molecular parameters, and reviews how rhizobia, a diverse group of α and β-proteobacteria bacteria, and legume species interact and are responsible for symbiotic biological nitrogen fixation.

Carmine Crecchio
Editor

 *diversity*

Editorial

Genetic Diversity of Soil Bacteria

Carmine Crecchio

Department of Soil, Plant and Food Sciences, University of Bari Aldo Moro, Via Amendola 165/a, 70126 Bari, Italy; carmine.crecchio@uniba.it

Received: 24 October 2020; Accepted: 28 October 2020; Published: 29 October 2020

Abstract: The Special Issue "Genetic Diversity of Soil Bacterial Communities" collected research and review articles addressing some relevant and unclear aspects of the composition and functioning of bacterial communities in rich or marginal agricultural soils, in field trials as well as in laboratory-scale experiments, at different latitudes and under different types of management.

Keywords: soil bacteria diversity; bacteriobiome; soil fertility

---

Soil has been defined as a "black box" because of its complexity and of the difficulties that scientists have faced in recent decades to unravel the composition and, moreover, the functions of its biotic and abiotic components. Soil has been recognized to play a key role in the biosphere, as most of the carbon and nutrient fluxes occur in its top 10 cm profile, and it is a habitat that is tremendously rich in species that contribute to plant growth and health that, in turn, may affect human beings [1].

Among the biotic components, the soil microbiome, mainly bacteria and at a lower scale fungi, plays an important role as a driver of the living soil, being responsible for the main biogeochemical cycles involved in the transformation of nutrients and their flows from soil to plants [2]. On the other hand, the soil microbiome is strongly affected by anthropogenic activities such as intensive agriculture, different soil management approaches, and environmental contamination [3,4].

So, it is very important to deeply investigate the structure and functioning of whole microbial communities as well as to characterize single species, i.e., those potentially useful in agriculture as promoters of plant nutrition and protection. The papers in this Special Issue are an example of how many different aspects should be investigated by biochemical, microbiological and molecular approaches, and reveal how varied and still not completely understood this topic is.

A research article investigated how various fodder and lawn grass species and their functional forms modify the soil bacteriobiome, as determined by 16S rDNA sequencing up to the genera level, as well as the main enzyme activities of microbial origin [5].

A similar approach was used to investigate the relevance of cultivation regimes of three crop species, leading the authors to conclude that both species and management contribute to the modification of the rhizosphere soil microbiome as well as reliable biochemical indicators related to soil fertility [6].

Yaghoubi Khanghahi and colleagues investigated the influence of long-term land use intensities on: (i) some microbiological and biochemical parameters, all contributing to the calculation of the Biological Fertility Index (BFI); (ii) total bacterial quantification in soil, determining the rRNA gene copy number by qPCR; (iii) α diversity and community composition [7].

A research paper by Marques and colleagues reports that sweet potato genotypes and their growth stage influence rhizosphere bacterial composition, in particular that of microbial groups involved in phosphate mineralization and nitrogen fixation, both playing a fundamental role in plant nutrition processes [8].

The Special Issue also publishes a review by Moura and colleagues, reporting the state of the art of the symbiotic nitrogen fixation, deepening, in particular, the interactions existing between rhizobia and legume species, and how such association provides benefits not only in the efficiency of nitrogen use, but also in soil carbon sequestration, stabilization of soil organic matter, soil penetration resistance and soil fertility in tropical environments [9].

We expect the above mentioned articles will be of interest to the scientific community and complement ongoing global efforts to reveal secrets of the "black box", soil and its bacterial inhabitants.

Funding: This research received no external funding.

Conflicts of Interest: The author declares no conflict of interest.

References

1. Nannipieri, P.; Ascher, J.; Ceccherini, M.T.; Landi, L.; Pietramellara, G.; Renella, G. Microbial diversity and soil functions. *Eur. J. Soil Sci.* **2017**, *68*, 12. [CrossRef]
2. Tate, R.L., III. *Soil Microbiology*; Wiley: New York, NY, USA, 1995.
3. Hendgen, M.; Hoppe, B.; Doring, J.; Friedel, M.; Kauer, R.; Frisch, M.; Dahl, A.; Kellner, H. Effects of different management regimens on microbial biodiversity in vineyard soils. *Sci. Rep.* **2018**, *8*, 9393. [CrossRef] [PubMed]
4. Kennedy, A.C.; Smith, K.L. Soil microbial diversity and sustainability of agricultural soils. *Plant Soil* **1995**, *170*, 75. [CrossRef]
5. Borowik, A.; Wyszkowska, J.; Kucharski, J. Impact of Various Grass Species on Soil Bacteriobiome. *Diversity* **2020**, *12*, 212. [CrossRef]
6. Wyszkowska, J.; Borowik, A.; Olszewski, J.; Kucharski, J. Soil Bacterial Community and Soil Enzyme Activity Depending on the Cultivation of *Triticum aestivum*, *Brassica napus*, and *Pisum sativum* ssp. *arvense*. *Diversity* **2019**, *11*, 246. [CrossRef]
7. Yaghoubi Khanghahi, M.; Murgese, P.; Strafella, S.; Crecchio, C. Soil Biological Fertility and Bacterial Community Response to Land Use Intensity: A Case Study in the Mediterranean Area. *Diversity* **2019**, *11*, 211. [CrossRef]
8. Marques, J.M.; Mateus, J.M.; da Silva, T.F.; Couto, C.R.A.; Blank, A.F.; Seldin, L. Nitrogen Fixing and Phosphate Mineralizing Bacterial Communities in Sweet Potato Rhizosphere Show a Genotype-Dependent Distribution. *Diversity* **2019**, *11*, 231. [CrossRef]
9. Moura, E.G.; Carvalho, C.S.; Bucher, C.P.C.; Souza, J.L.B.; Aguiar, A.C.F.; Ferraz Junior, A.S.L.; Bucher, C.A.; Coelho, K.P. Diversity of Rhizobia and Importance of Their Interactions with Legume Trees for Feasibility and Sustainability of the Tropical Agrosystems. *Diversity* **2020**, *12*, 206. [CrossRef]

Publisher's Note: MDPI stays neutral with regard to jurisdictional claims in published maps and institutional affiliations.

Article

Impact of Various Grass Species on Soil Bacteriobiome

Agata Borowik, Jadwiga Wyszkowska * and Jan Kucharski

Department of Microbiology, University of Warmia and Mazury in Olsztyn, 10-727 Olsztyn, Poland;
agata.borowik@uwm.edu.pl (A.B.); jan.kucharski@uwm.edu.pl (J.K.)
* Correspondence: jadwiga.wyszkowska@uwm.edu.pl

Received: 2 April 2020; Accepted: 22 May 2020; Published: 26 May 2020

Abstract: Today, various grass species are important not only in animal feeding but, increasingly often, also in energetics and, due to esthetic and cultural values, in landscape architecture. Therefore, it is essential to establish the roles various grass species and their functional forms play in modifying soil bacteriobiome and enzymatic activity. To this end, a pot experiment was conducted to examine effects of various fodder grass and lawn grass species on the bacteriobiome and biochemical properties of soil. Nonsown soil served as the control for data interpretation. Analyses were carried out with standard and metagenomic methods. The intensity of effects elicited by grasses depended on both their species and functional form. More favorable living conditions promoting the development of soil bacteria and, thereby, enzymatic activity were offered by fodder than by lawn grass species. Among the fodder grasses, the greatest bacteriobiome diversity was caused by sowing the soil with *Phleum pratense* (Pp), whereas among lawn grasses in the soil sown with *Poa pratensis* (Pr). Among the fodder grasses, the highest enzymatic activity was determined in the soil sown with *Lolium* x *hybridum* Hausskn (Lh), and among the lawn grasses—in the soil sown with *Lolium perenne*. Sowing the soil with grasses caused the succession of a population of bacterial communities from r strategy to k strategy.

Keywords: fodder grasses; lawn grasses; soil bacteria; soil enzymes

1. Introduction

Interactions between soil, plants, and soil microbiome are complex in character and require extended research. Determination of changes in soil stability and identification of associations between microbiological diversity of soil and plants occurring in agricultural ecosystems are difficult because they are affected by plant root secretions [1–4], climatic changes [5–7], and various pollutants [2,7]. As the key component of life on Earth, soil is capable of meeting most of plant demands. Its traits, including abundance of nutrients, productivity, and fertility, are a measure of the strength of plant growth and crop yield [8–11]. Plant productivity largely depends on soil culture [9], count of soil bacteria and fungi colonizing the rhizosphere [12,13], count of epiphytic microorganisms occurring on the surface of plants and endophytic ones colonizing their tissues [14], presence of pathogens [15,16], humus content [17,18], soil pH [19], water-air balance [20–22], soil fraction size [23,24] as well as the microbiological and biochemical activity of soil [10,25–27].

For agricultural sustainability, 176 cultivars have been shortlisted by the Research Center for Cultivar Testing (Słupia Wielka, Poland, 52.227° N 17.218° E) of which 19 are grasses species of monocotyledonous flowering plants from the *Poaceae* (*Gramineae*) family, commonly known as grasses, have been used in contemporary agriculture: x*Festulolium* Asch. & Graebn., *Festuca rubra* L., *Festuca pratensis* Huds., *Festuca filiformis* Pourr., *Festuca ovina* L., *Festuca trachyphylla* (Hack.) Krajina, *Festuca arundinacea* Schreber, *Dactylis glomerata* L., *Agrostis gigantea* Roth, *Agrostis capillaris* L.,

Agrostis stolonifera L., *Arrhenatherum elatius* (L.) P. Beauv.ex J. Presl & C. Presl, *Bromus catharticus* Vahl, *Phleum pratense* L., *Poa pratensis* L., *Poa trivialis* L., *Lolium x hybridum* Hausskn, *Lolium perenne* L., and *Lolium multiflorum* Lam. Pursuant to EU regulations, all cultivars submitted to the national register are evaluated for distinctness, uniformity and stability (DUS), whereas crops are additionally evaluated for their value for cultivation and use [28]. Grasses from the *Poaceae* family, i.e., from the family of monocotyledoneous flowering plants, represent one of the most important and the most abundant group of plants on the entire Earth. This family includes crops and monocotyledoneous fodder plants. These plants constitute the source of feed to both, wild and domesticated animals. They possess therapeutic and health-promoting properties, and are able to adapt to various climatic zones and various habitats. The form assemblages of savannas, steppes, prairies and pampas, as well as lowland, mountain, and arctic meadows. They have been accompanying man for years. They have been and are used most often in animal feeding due to their high nutritive value, resulting from the chemical composition of plants. Grasses are rich in dietary fiber digestible protein, minerals, and vitamins [29]. According to Peeters [30], grasses have a higher nutritive value for animals than fodder beet. They represent complete feeds rich in organic and mineral compounds. Their leaves and stems may be easily ingested by animals and effectively digested by microorganisms colonizing their rumens. In addition, they are valuable energetic feed.

For sustainability, modern agricultural practices need to include every effort not to deplete the soil's organic matter, because the use of chemicals together with intensive cultivation can lead to soil sterilization and microbiological imbalance [11,31,32]. The development of soil edaphon is at risk of the impairment of decomposition and humification processes due to organic matter accumulating in the soil [33].

Soil microorganisms and enzymes take part in the mineralization of organic substances [34–39], in retention of heavy metals [40–42], and in degradation of plant protection agents [43–45] and polycyclic aromatic hydrocarbons (PAHs) [34,36,46]. They are the driving force of the geochemical cycle of elements, and participate in transformations of simple and complex organic compounds [47]. Diversity of microorganisms influences the functioning of ecosystems, biological homeostasis as well as chemical and physical properties of soil, and by this means determines its productivity [35,48,49].

Microorganisms that colonize soil and other environments synthesize intra- and extracellular enzymes indispensable for depolymerization and hydrolysis of organic macromolecules which serve as sources of carbon and energy [50]. Determination of enzymatic activity of soil is essential to the understanding of the functional dynamics of a soil ecosystem. According to Moeskops et al. [51], Zhan et al. [52], and Knight and Dick [53], it is also a good indicator of the biological status of soil because the activity of enzymes from the class of oxidoreductases (dehydrogenases or catalase) is strictly responsible for respiration of microorganisms in the soil. A reliable indicator of changes undergoing in the soil is also the activity of urease. Although this is an extracellular enzyme related to a lesser extent with the condition of microorganisms, it is highly sensitive to various xenobiotics [52]. In turn, β-glucosidade is responsible for cellulose transformation to glucose [53], while phosphatases—for transformations of phosphorus compounds [54], and, inter alia, arylsulfatase—for the metabolism of organic sulfur [55]. It can therefore be concluded that the geochemical transformations proceeding in the soil are strongly associated with its biological activity.

Due to the small amount of research into the effects of grasses on soil biodiversity, research was undertaken to compare (1) the soil bacteriobiome of six grasses species (three fodder and three lawn grasses); (2) the effect of grasses on colony development and ecophysiological diversity index of soil bacteria; (3) the grass yield of fodder and lawn grasses; and (4) the enzymatic activities of soil with grasses and without grass.

2. Materials and Methods

2.1. Soil Characteristics

The experiment was conducted with eutric cambisol soil sampled from the topsoil, from a depth of 0 to 20 cm of the arable lands from the Olsztyn Lake District situated in the northeast of Poland (NE Poland, 53.7161° N 20.4167° E). It contained 74.93% of the sand fraction, 22.85% of the silt fraction, and 2.22% of the clay fraction. In terms of fraction size, this was loamy sand [56]. The physicochemical and chemical properties of soil are presented in Table 1. They were conducted according to the procedures presented in the manuscript by Borowik et al. [57].

Table 1. Physicochemical and chemical soil properties.

pH_{KCl}	HAC	EBC	CEC	BS %	Content		Available Forms			Interchangeable Forms			
	mmol (+) kg^{-1} d.m. of Soil				N_{total}	C_{total}	P	K	Mg	K	Ca	Na	Mg
					g kg^{-1} d.m. of Soil		mg kg^{-1} d.m. of Soil						
6.70	11.40	49.00	60.40	81.10	0.62	9.30	93.68	141.10	42.00	156.00	623.50	40.00	59.50

HAC—hydrolytic activity, EBC—exchangeable base cations, CEC—cation exchange capacity, BS—base saturation, d.m.—dry matter.

2.2. Plant Characteristics

The study focused on plants having a well-developed root system, including three species of lawn grasses and three species of fodder grasses (Table 2).

Table 2. Characteristics of grasses used in the study.

Grasses Kind	Common Name	Botanical Name	Abbreviation	Variety	Photosynthesis Kind
Fodder	Hybrid ryegrass	*Lolium x hybridum* Hausskn	Lh	Gala	C3
	Tall fescue	*Festuca arundinacea*	Fa	Rahela	C3
	Timothy	*Phleum pratense*	Pp	Kaba	C3
Lawn	Perennial ryegrass	*Lolium perenne*	Lp	Bajka	C3
	Smooth-stalked meadowgrass	*Poa pratensis*	Pr	Sójka	C3
	Red rescue	*Festuca rubra*	Fr	Dark	C3

2.3. Experimental Design

The study was conducted in a pot experiment, at the teaching-experimental station of the University of Warmia and Mazury in Olsztyn (NE Poland, 53.760° N 20.454° E). It was accomplished in two series: with nonsown (without grasses) soil and with soil sown with the selected grass species. The experiment (pots sown with six different species of grass and soil without grasses) was performed in four replications in 10 dm^3 Kick–Brauckman pots, each filled with 9 kg of soil. Before the experiment had been established, the soil was sieved through a screen with mesh diameter of 5 mm, then thoroughly mixed, weighed into 9-kg portions, carefully mixed with mineral fertilizers, and poured into the pots. With soil sowing, 22 seeds were sown to each pot. The same mineral fertilization was applied for all grass species and control soil (not sown with grasses). The pre-sowing fertilization included, in mg kg^{-1} soil d.m. (dry matter): N—80, P—20, K—40, and Mg—10, whereas, after the harvest of the first and the second re-growth, the plants were additionally fertilized with nitrogen in the amount of 40 mg N kg^{-1} soil d.m.. All grass species emerged evenly and at the same time. After emergence, 20 plants were left in each pot. The experiment spanned for 105 days. Within this period, soil humidity was kept at a level of 50% of the maximum water capacity. The grasses were cut three times. Each time, the biomass of aerial parts was determined. In the last term of cutting (day 105 of experiment). Plants were removed from the pots and then the soil from each pot was mixed thoroughly.

2.4. Determination of Bacterial Count and Activity of Soil Enzymes

In soil samples, microbiological and biochemical analyses were carried out using standard methods, which are given in Table 3. These methods are described in detail in the manuscript of Borowik et al. [57] and Wyszkowska et al. [58]. Analyses were performed in four replications.

Table 3. Parameters for determining the number of organotrophic bacteria and actinobacteria, calculating colony development index and ecophysiological diversity index, and determining enzyme activity.

Tested Feature	Medium/Formula/Substrat	References
	Medium	
Organotrophic bacteria (Org)	peptone 1.0 g, yeast extract 1.0 g, $(NH_4)_2SO_4$ 0.5 g, $CaCl_2$, K_2HPO_4 0.4 g, $MgCl_2$ 0,2 g, $MgSO_4$ $7H_2O$ 0.5 g, salt Mo 0.03 g, $FeCl_2$ 0.01 g, agar 20.0 g, soil extract 250 cm^3, distilled water 750 cm^3, pH 6.6–7.0	[59] [58]
Actinobacteria (Act)	soluble starch 10.0 g; casein 0.3 g; KNO_3 2.0 g; NaCl 2.0 g; K_2HPO_4 2.0 g; $MgSO_4 \cdot 7H_2O$ 0.05 g; $CaCO_3$ 0.02 g; $FeSO_4$ 0.01 g; agar 20.0 g; H_2O 1 dm^3; 50 cm^3 aqueous solution of nystatin 0.05%; 50 cm^3 aqueous solution of actidione 0.05%; pH 7.0	[60] [58]
	Formula	
Colony development index (CD)	CD = [N1/1 + N2/2 + N3/3 + … + N10/10] × 100, where: N1, N2, N3,…, N10—the sum of ratios of the number of colonies of microorganisms identified in particular days (1, 2, 3,…, 10) to the total number of colonies identified throughout the study period	[61]
Ecophysiological diversity index (EP)	EP = −Σ(pi·log10 pi), where: pi—the ratio of the number of colonies of microorganisms identified in particular days to the total number of colonies identified throughout the study period.	
	Substrat	
Dehydrogenases (Deh)	$C_{19}H_{15}ClN_4$	[62]
Catalase (Cat)	H_2O_2	[63]
Urease (Ure)	CON_2H_4	
β-glucosidase (Glu)	$C_{12}H_{15}NO_8$	
Acid phosphatase (Pac) Alkaline phosphatase (Pal)	$O_2NC_6H_4OP(O)(ONa)_2$ $6H_2O$	[64]
Aryosulphatase (Aryl)	$NO_2C_6H_4OSO_2OK$	

2.5. Metagenomic Analysis

Taxon of bacteria in soil samples was determined using analysis of the 16S rRNA encoding gene based on the hypervariable region V3–V4. Two primers were used for amplification: 1055F (5′-ATGGCTGTCGTCAGCT-3′) and 1392R (5′-ACGGGCGGTGTGTAC-3′). Polymerase chain reaction (PCR) was performed in real time in an Mx3000P thermocycler (Stratagene) and sequencing was in an Mx3000P thermocycler (Stratagene) at Genomed S.A. Warsaw, Poland.

2.6. Bioinformatic Analysis

The classification of bacteria was carried out with the QIIME package based on reference sequences database Greengenes v13_8 [65]. Sequences shorter than 1250 base pairs (bp), incomplete sequences and sequences containing more than 50 degenerated bases were omitted in the analysis and reference databases were prepared. The sequences were grouped in operational taxonomic units (OTUs).

2.7. Statistical Analysis

The Statistica 13.1 package [66] was used for statistical analyses. Homogenous groups were determined with the Tukey's test, at p = 0.05, and respective results were presented graphically using principal component analysis (PCA) and graphs categorized for dependent variables (category X) and the grouping variable (category Y). Using the analysis of variance (ANOVA), F and P values were calculated for all parameters tested (Table S1). Relative abundance of microorganisms in soil samples was visualized with the use of STAMP 2.1.3 software, using a two-way test for statistical hypotheses: G-test (w/Yates') + Fisher's and Asymptotic with CC confidence interval method [67]. Genomic data were presented in the circular system using Circos 0.68 package [68]. Visualization of relative abundance data was performed using sequences with contribution exceeding 1%. The read-outs below 1% were summed up with the other nonclassified ones in a sample. To determine bacterial diversity at the level of each taxonomic group, Shannon–Wiener (H) and Simpson (D) indices were calculated using all metagenomic data. In addition, in order to consider not only the role of individual grass species in soil bacteriobiome modification but also to emphasize functional types of grasses, the fodder grasses: *Lolium perenne* (Lp), *Festuca arundinacea* (Fa), and *Phleum pratense* (Pp) were grouped and marked as T1, whereas the lawn grasses: *Lolium perenne* L.× *hybridum* (Lh), *Poa pratensis* (Pr), and *Festuca rubra* (Fr)—as T2.

3. Results

3.1. Grass Yield

The growth and development of grasses were significantly affected by their species (Figure 1). Generally, regardless of cutting term and grass species, the yield of fodder grasses (T1) was higher by 71.68% on average than that of the lawn grasses (T2). Among the fodder grasses, in the first and third terms of harvest—the greatest biomass yield was obtained from *Lolium x hybridum* Hausskn (Lh), whereas in the second term—from *Festuca arundinacea* (Fa). In the case of lawn grasses, the best yield was produced by *Lolium perenne* (Lp) in all three terms of harvest. To sum up, regardless of the harvest date, the largest biomass among fodder grasses was obtained in the case of *Lolium x hybridum* Hausskn (Lh), and, among lawn grasses, *Lolium perenne* (Lp).

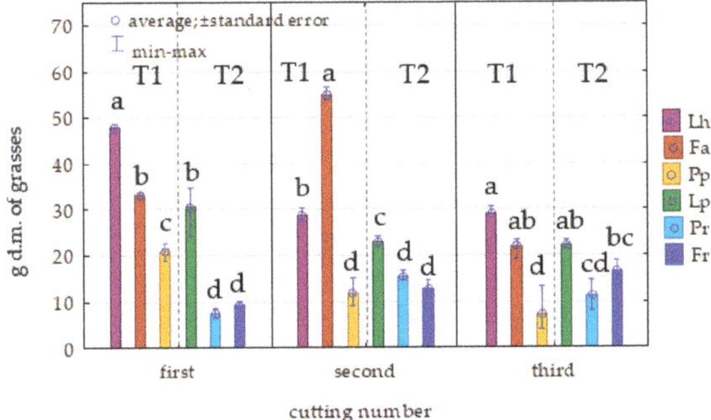

Figure 1. The yield of grasses in g dry matter (d.m.) per pot. T1—fodder grasses; T2—lawn grasses; homogeneous groups denoted with letters (a–d) were calculated separately for every cutting; Lh—*Lolium x hybridum* Hausskn; Fa—*Festuca arundinacea*; Pp—*Phleum pratense*, Lp—*Lolium perenne*, Pr—*Poa pratensis*; Fr—*Festuca rubra*.

3.2. Counts and Diversity of Soil Bacteria

Cultivation of grasses ambiguously modified soil bacteriobiome (Figure 2). Sowing the soil with *Lolium x hybridum* Hausskn (Lh) caused a significant increase in the population number of both organotrophs and actinobacteria, whereas cultivation of *Festuca arundinacea* (Fa), *Lolium perenne* (Lp), and *Festuca rubra* (Fr)—only in the population number of actinobacteria. In turn, sowing *Phleum pratense* (Pp), *Lolium perenne* (Lp), *Poa pratensis* (Pr), and *Festuca rubra* (Fr) on soils significantly reduced the proliferation of organotrophic bacteria. Regardless of grass species, but considering their functional character, it was demonstrated that the fodder grasses (T1) increased the count of organotrophic bacteria by 26.57% and that of actinobacteria by 156.49% compared to the control soil (not sown with grasses), whereas the lawn grasses (T2) increased the count of actinobacteria by 47.43% and decreased that of organotrophs by 37.91% compared to the control soil.

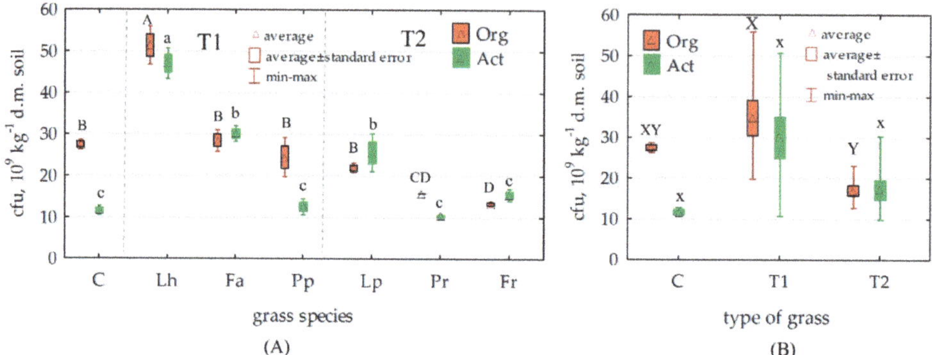

Figure 2. Count of soil organotrophic bacteria (Org) and actinobacteria (Act), cfu 10^9 kg^{-1} d.m. of soil: (**A**) depending on the species of grass, (**B**) depending on grass type (fodder or lawn). Homogeneous groups denoted were calculated separately for each microorganisms, groups denoted with letters (a–d) were calculated for the species of grass and groups denoted with letters (x–y) were calculated for the type of grass. C—unsown soil; Lh—*Lolium x hybridum* Hausskn; Fa—*Festuca arundinacea*; Pp—*Phleum pratense*, Lp—*Lolium perenne*, Pr—*Poa pratensis*; Fr—*Festuca rubra*.

The positive impact of the fodder grasses on the proliferation of soil microorganisms was not reflected in their ecophysiological diversity index (EP), because cultivation of Lh, Fa, and Pp not only did not increase the EP index of organotrophs but decreased its value by 7.9% on average, and also did not change the EP index of actinobacteria (Figure 3). Also the cultivation of the lawn grasses caused insignificant changes in the value of EP index of organotrophic bacteria, whereas Lp and Fr had the same effect also on actinobacteria. Only Pr decreased EP index of actinobacteria by 8.14%.

Sowing the soil with both fodder and lawn grasses caused a significant decrease in the values of the colony development index (CD) calculated for the organotrophic bacteria (Figure 4). A decrease in CD value calculated for organotrophic bacteria ranged from 27.80% (Fa) to 36.88% (Lh) in the case of fodder grass species, and from 30.02% (Fr) to 37.73% (Pr) in the case of lawn grass species. A lesser decrease in CD value was observed in the case of actinobacteria, i.e., from 3.97% (Fa) to 16.26% (Pp) in the soils used to cultivate fodder grasses, and from 6.23% (Fr) to 16.93% (Pr) in the soils sown with lawn grasses.

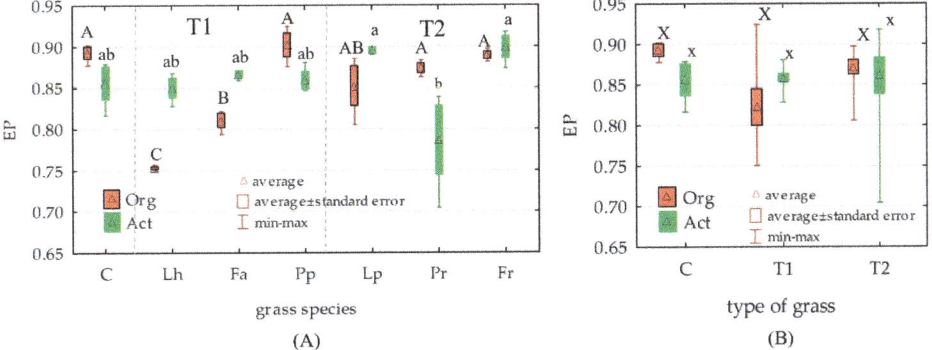

Figure 3. Ecophysiological diversity index (EP) of organotrophic bacteria (Org) and actinobacteria (Act): (**A**) depending on the species of grass, (**B**) depending on grass type (fodder or lawn). Homogeneous groups denoted were calculated separately for each microorganisms, groups denoted with letters (a–c) were calculated for the species of grass and groups denoted with letters (x) were calculated for the type of grass. C—unsown soil; Lh—*Lolium* x *hybridum* Hausskn; Fa—*Festuca arundinacea*; Pp—*Phleum pratense*, Lp—*Lolium perenne*, Pr—*Poa pratensis*; Fr—*Festuca rubra*.

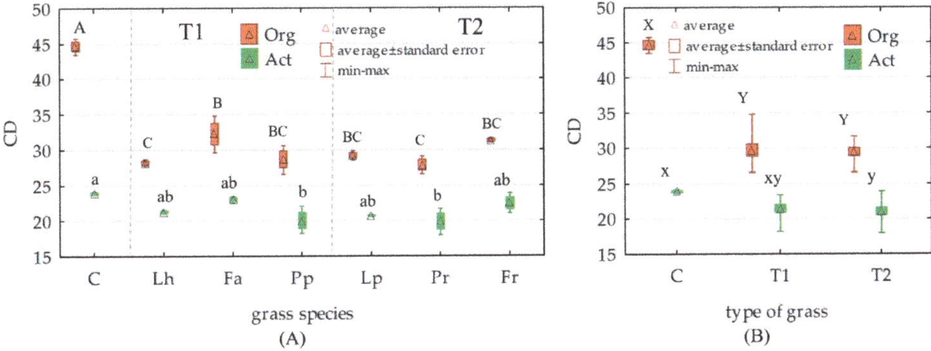

Figure 4. Colony development index (CD) of organotrophic bacteria (Org) and actinobacteria (Act): (**A**) depending on the species of grass, (**B**) depending on grass type (fodder or lawn). Homogeneous groups denoted were calculated separately for each microorganisms, groups denoted with letters (a–d) were calculated for the species of grass and groups denoted with letters (x–y) were calculated for the type of grass. C—unsown soil; Lh—*Lolium* x *hybridum* Hausskn; Fa—*Festuca arundinacea*; Pp—*Phleum pratense*, Lp—*Lolium perenne*, Pr—*Poa pratensis*; Fr—*Festuca rubra*.

At all plots, the prevailing phyla included *Proteobacteria* and *Actinobacteria* (Figure S1). In the control soil, *Proteobacteria* accounted for 28.78% of total bacteria, whereas this fraction was 38.56% in the soil sown with fodder grasses and 32.93% in that sown with lawn grasses. In the control soil, *Actinobacteria* accounted for 23.55%, in the soil sown with fodder grasses, for 19.77%, and, in the soil sown with lawn grasses, for 21.32% of total bacteria. The OTU number of *Proteobacteria* in the soil sown with fodder and lawn grasses was higher by 9.8% than in the control soil, whereas the OTU number of *Actinobacteria* decreased by 3.4% in the soil sown with fodder grasses and by 2.9% in soil sown with lawn grasses, compared to the control soil.

Apart from *Proteobacteria* and *Actinobacteria*, taxa identified at the phylum level included: *Acidobacteria, Chloroflexi, Gemmatimonadetes, Firmicutes* and *Planctomycetes, Bacteroidetes, Verrucomicrobia, Cyanobacteria*, as well as OD1 and TM7 (Figure 5A,B). The OTU number of bacteria classified as 'others' reached 4.3% in the control soil, 3.38% in the soil sown with fodder grasses, and 3.80% in the soil sown with lawn grasses. Cultivation of all grass species facilitated the proliferation of *Proteobacteria*, *Bacteroidetes*, and *Verrucomicrobia* bacteria, which resulted in higher OTU numbers of these bacteria compared to the control soil. The OTU number of *Actinobacteria* increased only in the sample of soil sown with *Lolium* x *hybridum* Hausskn (Lh) and *Lolium perenne* (Lp); the OTU number of *Chloroflexi* increased in the soils sown with *Lolium perenne* (Lp), *Poa pratensis* (Pr), *and Festuca rubra* (Fr); the OTU number of Firmicutes increased in the soil sown with *Poa pratensis* (Pr), whereas the OTU number of *Planctomycetes* rose in the soil sown with *Poa pratensis* (Pr), and *Festuca arundinacea* (Fa).

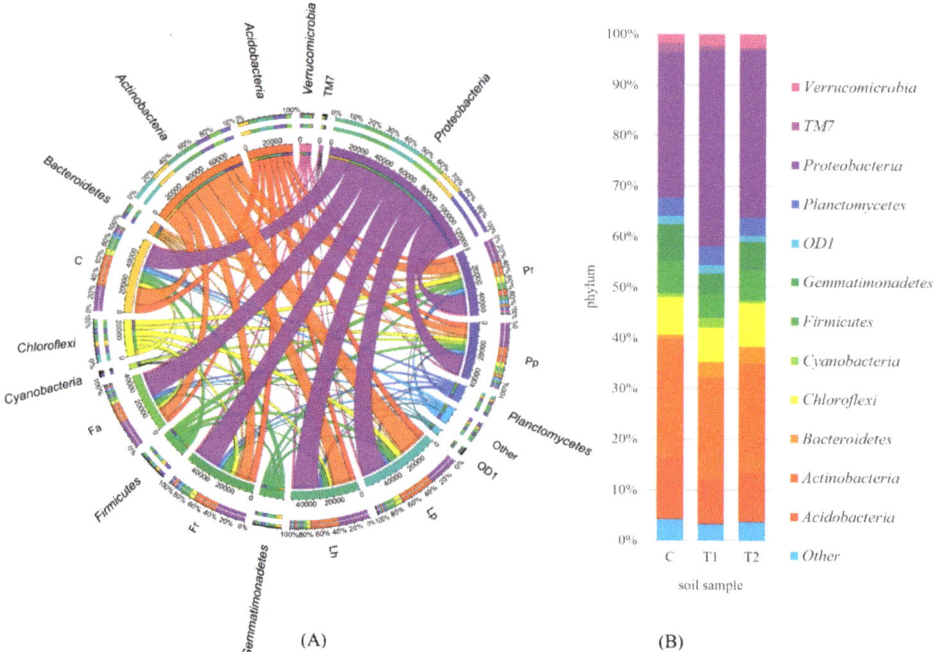

Figure 5. Bacterial communities at the phylum level (**A**) depending on the species of grass, (**B**) depending on grass type (fodder or lawn). Abundances <1% are gathered into the category "other". C—unsown soil; T1—average bacteria abundance in soils sown with fodder grasses; T2—average bacteria abundance in soils sown with lawn grasses. Lh—*Lolium* x *hybridum* Hausskn; Fa—*Festuca arundinacea*; Pp—*Phleum pratense*, Lp—*Lolium perenne*, Pr—*Poa pratensis*; Fr—*Festuca rubra*.

Both, in the control soil and soils sown with grasses, the prevailing class of bacteria was *Alphaproteobacteria*, which in the control soil accounted for 19.32% of total bacteria, in the soil sown with fodder grasses—for 19.99%, and in the soil sown with lawn grasses—for 16.50% (Figure S2). The second prevailing class was *Actinobacteria*, which in the structure of all bacterial classes represented 12.98% in the control soil, 13.88% in the soil used to cultivate fodder grasses, and 13.98% in the soil used to grow lawn grasses. Sowing the soils with fodder grasses caused the greatest changes in the abundance of bacterial classes: *Gammaproteobacteria*, *Betaproteobacteria*, and *Acidobacteria*-6, whose OTU numbers increased by 6.37%, 2.64%, and 1.99%, respectively, compared to the control soil.

Also, sowing the soil with lawn grasses increased OTU numbers of these bacteria in the range from 3.56% (*Gammaproteobacteria*) to 1.92% (*Acidobacteria*-6). Differences in changes in the bacterial structure were also noticeable between the rhizospheres of the fodder and lawn grasses. The OTU number of *Alphaproteobacteria* in the soils sown with fodder grasses was higher by 3.49% and that of *Gammaproteobacteria* by 2.82% than in the soils sown with lawn grasses. Regardless of grass species and functional designation, apart from the two prevailing classes *Alphaproteobacteria* and *Actinobacteria,* all soils contained also (in a descending order of OTUs): *Betaproteobacteria, Thermoleophilia, Gammaproteobacteria, Bacilli, Acidobacteria-6, Gemmatimonadetes, Deltaproteobacteria, Planctomycetia, Solibacteres, Acidimicrobiia, Acidobacteriia,* Gemm-1, C0119, Ellin6529, *Chloracidobacteria, Clostridia, Phycisphaerae, Saprospirae, Ktedonobacteria, Pedosphaerae,* ZB2, *Thermomicrobia, Spartobacteria, Sphingobacteriia, Verrucomicrobiae, Chloroflexi, Chloroplast,* TM7-1, *Flavobacteriia,* and *Nostocophycideae* (Figure 6A,B).

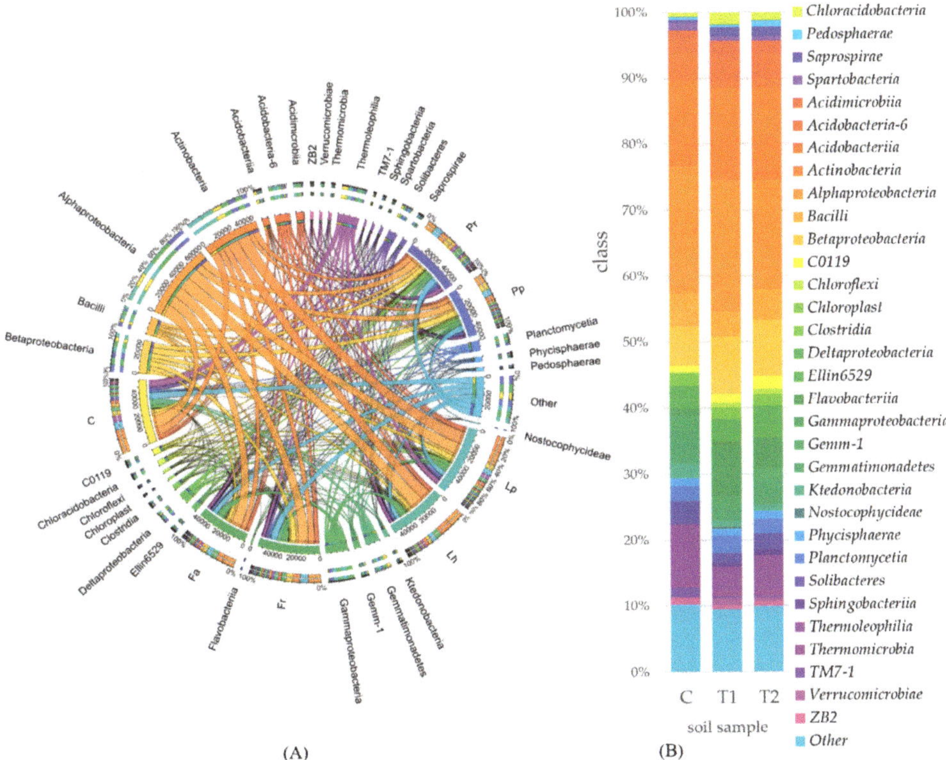

Figure 6. Bacterial communities at the class level (**A**) depending on the species of grass, (**B**) depending on grass type (fodder or lawn). Abundances <1% are gathered into the category "other". C—unsown soil; T1—average bacteria abundance in soils sown with fodder grasses; T2—average bacteria abundance in soils sown with lawn grasses. Lh—*Lolium* x *hybridum* Hausskn; Fa—*Festuca arundinacea*; Pp—*Phleum pratense*, Lp—*Lolium perenne*, Pr—*Poa pratensis*; Fr—*Festuca rubra*.

The effect of grasses on soil bacteriobiome was also noticeable at the order level. Among the 37 identified orders with OTU numbers above 1%, the greatest abundance was demonstrated for bacteria classified to *Actinomycetales*, *Sphingomonadales*, and *Rhizobiales* (Figure S3). When comparing effects of various functional types of grasses, it was found that sowing the soils with fodder grasses increased OTU numbers of *Xanthomonadales* by 4.67%, *Sphingomonadales* by 3.18%, *Burkholderiales* by 2.41%, *Solibacterales* by 2.04%, and *Actinomycetales* by 1.12%, whereas sowing the soils with lawn grasses increased OTU numbers of *Burkholderiales* by 2.60%, *Xanthomonadales* by 1.85%, *Alteromonadales* by 1.39%, *Actinomycetales* by 1.23, and *Rhizobiales* by 1.02%, compared to the nonsown control soil.

The highest OTU number of *Actinomycetales* bacteria was determined in the soils sown with *Lolium* x *hybridum* Hausskn (Lh) and *Lolium perenne* (Lp); whereas that of *Sphingomonadales*—in the soils sown with *Lolium* x *hybridum* Hausskn (Lh), *Festuca arundinacea* (Fa), *Festuca rubra* (Fr), and *Lolium perenne*

(Lp) (Figure 7A,B). Higher OTU numbers were also determined in the soils with growing grasses than in the control soil for the bacteria from the following orders: *Burkholderiales, Xanthomonadales, Saprospirales,* and *Sphingobacteriales*.

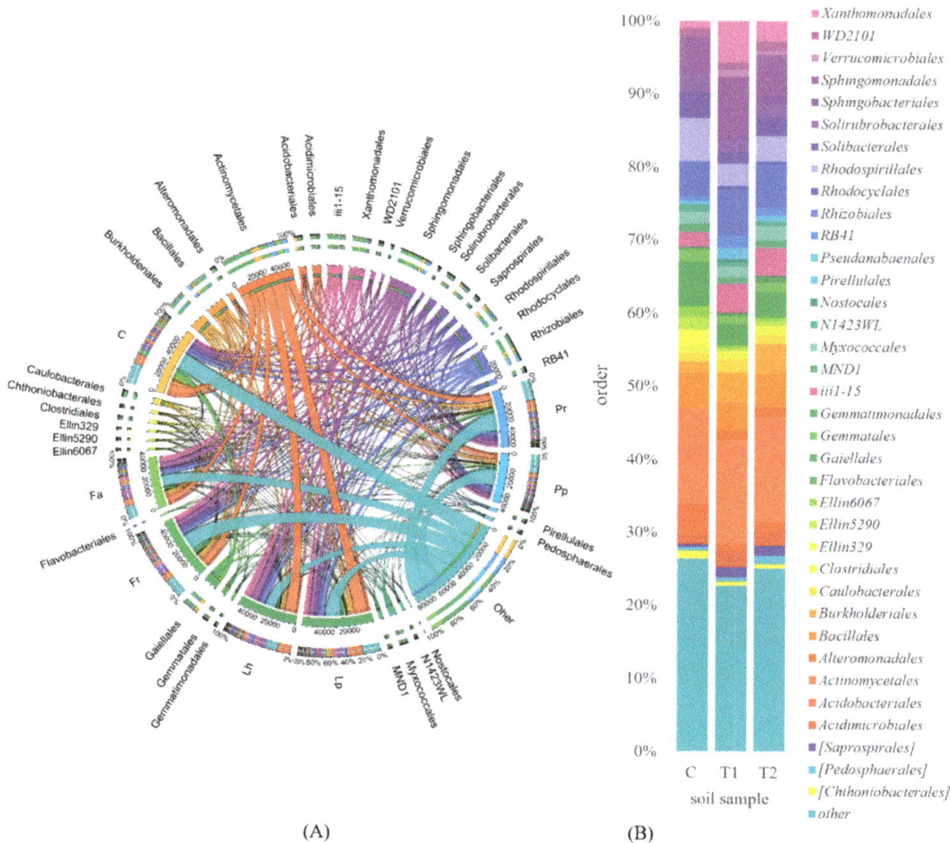

Figure 7. Bacterial communities at the order level (**A**) depending on the species of grass, (**B**) depending on grass type (fodder or lawn). Abundances <1% are gathered into the category "other". C—unsown soil; T1—average bacteria abundance in soils sown with fodder grasses; T2—average bacteria abundance in soils sown with lawn grasses. Lh—*Lolium* x *hybridum* Hausskn; Fa—*Festuca arundinacea*; Pp—*Phleum pratense*, Lp—*Lolium perenne*, Pr—*Poa pratensis*; Fr—*Festuca rubra*.

Differences in the abundance of bacterial populations were also observed at the family level. Compared to the control soil, the greatest changes in the structure of bacteria classified to families, after soil sowing with both fodder and lawn grasses, occurred in the families of *Gaiellaceae* (decrease by 3.16% and 2.57%, respectively), *Koribacteraceae* (decrease by 2.77% and 2.54%), *Intrasporangiaceae* (increase by 2.14% and 1.38%), *Xanthomonadaceae* (increase by 3.54% and 1.13%), and *Comamonadaceae* (increase by 2.61% and 2.43) (Figure S4).

When comparing effects of individual grass species on OTU number of bacteria classified to families, it can be concluded that they were inexplicit (Figure 8A,B). All species increased OTU numbers of the following families: *Intrasporangiaceae, Bradyrhizobiaceae, Xanthomonadaceae, Sinobacteraceae, Comamonadaceae, Pirellulaceae, Chitinophagaceae, Ellin5301, Nostocaceae,* and *Rhodocyclaceae,* but decreased OTU numbers of *Gaiellaceae, Rhodospirillaceae, Koribacteraceae, Solibacteraceae, Acetobacteraceae, Pseudonocardiaceae, Paenibacillaceae, Frankiaceae,* and *Chthoniobacteraceae.* Noteworthy is the fact that sowing the grasses onto soils resulted in the appearance of families: *Alteromonadaceae, Methylophilaceae, Flavobacteriaceae,* and *Verrucomicrobiaceae,* that were not identified in the control soil.

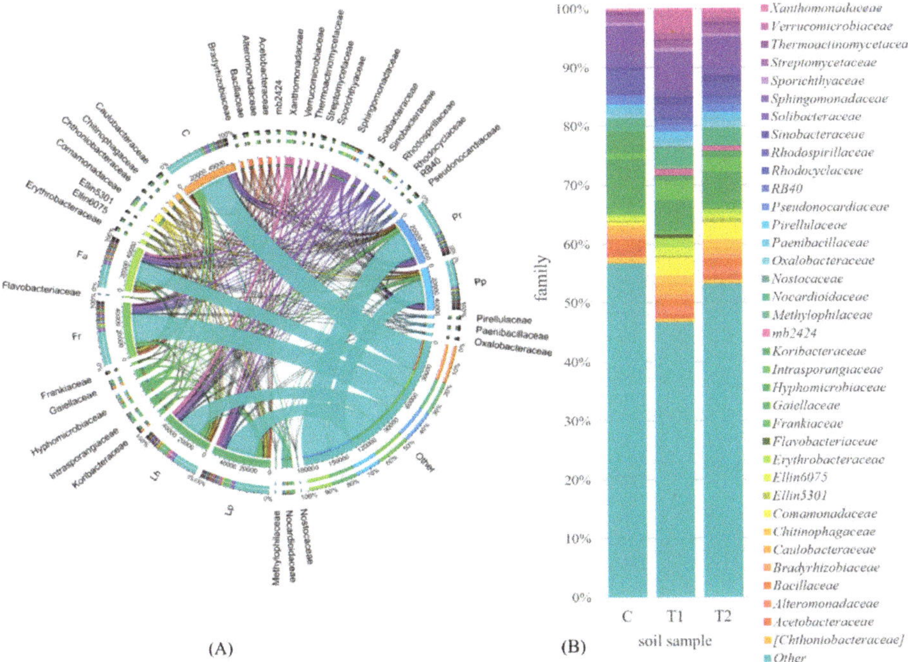

Figure 8. Bacterial communities at the family level (**A**) depending on the species of grass, (**B**) depending on grass type (fodder or lawn). Abundances <1% are gathered into the category "other". C—unsown soil; T1—average bacteria abundance in soils sown with fodder grasses; T2—average bacteria abundance in soils sown with lawn grasses. Lh—*Lolium* x *hybridum* Hausskn; Fa—*Festuca arundinacea*; Pp—*Phleum pratense*, Lp—*Lolium perenne*, Pr—*Poa pratensis*; Fr—*Festuca rubra*.

Considering OTU numbers above 1%, only 11 bacterial genus were identified in the control soil not sown with grasses, whereas 17 genus in the soil sown with Pp and Pr, 16 genus in the soil sown with Fr and Lp, 15 genus in the soil sown with Fa, and 14 genus in the soil sown with Lh (Figure 9). The contribution of the identified bacteria in the genus structure ranged from 13.86% in the soil sown with Pr to 20.34% in the soil sown with Lh. The genus *Kaistobacter* was found to predominate on all plots. Regarding OTU number, it was followed by *Rhodoplanes* in the control soil and soil sown with Pr, by *Terracoccus* in the soil sown with Lh and Lp, by HB2-32-21 in the soil sown with Fa and Fr, and by *Flavobacterium* in the soil sown with Pp. Compared to the soils overgrown with grasses, no OTUs of the following genera were identified in the control soil: *Arenimonas, Dechloromonas, Flavobacterium, Methylotenera,* and *Mycoplana.*

The analysis of values of Shannon and Simpson diversity indices points to a richer microbiome of the soils sown with grasses compared to the control soil (Table 4). The greatest abundance among the fodder grasses was found in the rhizosphere of Pp, and, among the lawn grasses, in the rhizosphere of Pr.

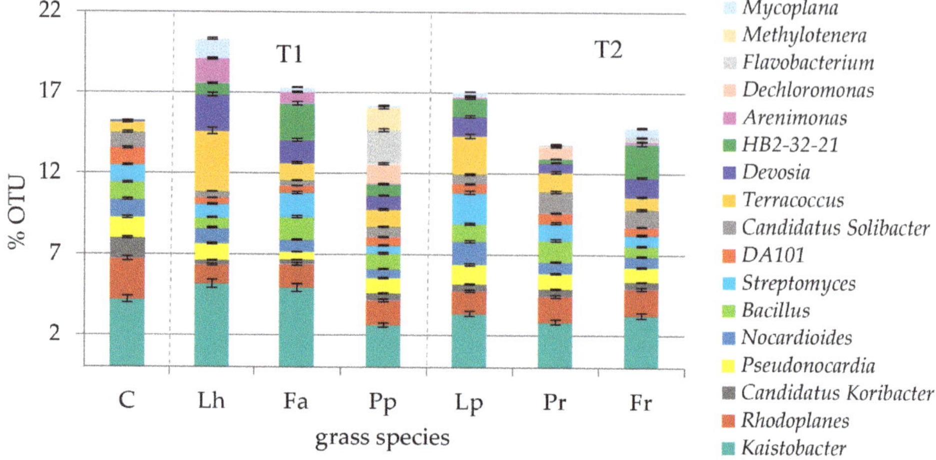

Figure 9. The operational taxonomic units (OTUs) structure of identified genus of bacteria in the total number of OTUs. C—unsown soil; T1—fodder grasses; T2—lawn grasses. Lh—*Lolium* x *hybridum* Hausskn; Fa—*Festuca arundinacea*; Pp—*Phleum pratense*, Lp—*Lolium perenne*, Pr—*Poa pratensis*; Fr—*Festuca rubra*.

Table 4. Shannon and Simpson indices calculated from abundance of OTU.

Taxon	C	Lh	Fa	Pp	Lp	Pr	Fr	C	T1	T2
				Shannon-Wiener index						
phylum	2.05 [ab]	1.86 [d]	1.92 [cd]	2.10 [ab]	1.90 [d]	2.12 [a]	2.01 [ac]	2.05 [x]	1.96 [x]	2.01 [x]
class	2.81 [bc]	2.67 [c]	2.80 [bc]	2.97 [a]	2.82 [abc]	2.96 [ab]	2.90 [ab]	2.81 [x]	2.81 [x]	2.89 [x]
order	2.77 [b]	2.75 [b]	2.90 [a]	2.91 [a]	2.81 [ab]	2.79 [ab]	2.93 [a]	2.77 [x]	2.86 [x]	2.84 [x]
family	2.00 [d]	2.49 [a]	2.43 [ab]	2.23 [c]	2.34 [bc]	2.04 [d]	2.26 [c]	2.00 [z]	2.38 [x]	2.21 [y]
genus	0.76 [c]	0.97 [a]	0.86 [b]	0.87 [b]	0.87 [b]	0.74 [c]	0.78 [c]	0.76 [z]	0.90 [x]	0.80 [y]
				Simpson index						
phylum	0.84 [ab]	0.78 [c]	0.78 [c]	0.83 [ab]	0.80 [b]	0.85 [a]	0.82 [ab]	0.84 [x]	0.80 [x]	0.82 [x]
class	0.92 [ab]	0.89 [b]	0.92 [ab]	0.93 [a]	0.92 [ab]	0.94 [a]	0.93 [a]	0.92 [x]	0.91 [x]	0.93 [x]
order	0.94 [ab]	0.91 [b]	0.94 [ab]	0.96 [a]	0.92 [b]	0.95 [a]	0.95 [a]	0.94 [x]	0.94 [x]	0.94 [x]
family	0.67 [ef]	0.82 [a]	0.78 [ab]	0.70 [de]	0.75 [bc]	0.66 [f]	0.72 [cd]	0.67 [y]	0.77 [x]	0.71 [y]
genus	0.28 [cd]	0.36 [a]	0.32 [b]	0.30 [c]	0.31 [b]	0.26 [d]	0.28 [cd]	0.28 [y]	0.33 [x]	0.28 [y]

Homogeneous groups denoted were calculated separately for each taxon groups denoted with letters (a–f) were calculated for the species of grass and groups denoted with letters (x–z) were calculated for the type of grass. C—unsown soil; Lh—*Lolium* x *hybridum* Hausskn; Fa—*Festuca arundinacea*; Pp—*Phleum pratense*; Lp—*Lolium perenne*, Pr—*Poa pratensis*; Fr—*Festuca rubra*.

3.3. Activity of Soil Enzymes

Study results demonstrated the highest enzymatic activity in the soil sown with Lh and Lp (Table 5, Figure 10), and the lowest one in the control soil. Activities of all enzymes were expressed in activity units per dry matter of 1 kg of soil within 1 h, and so the activity of dehydrogenases ranged from 10.292 µmol TFF (triphenyl formazan) in the soil sown with Lh to 0.654 µmol TFF with Pr; that of catalase, from 0.205 mol O_2 in the soil sown with Pr to 0.091 mol O_2 in the control soil; that of urease, from 0.880 mmol N-NH_4 in the soil sown with Pp to 0.302 in the control soil; that of acid phosphatase, from 1.302 mmol PNP (p-nitrophenyl) in the soil sown with Lp to 0.785 mmol PNP in the control soil; that of alkaline phosphatase, from 0.397 mmol PNP in the soil sown with Lh to 0.138 mmol PNP in the soil sown with Fr; that of β-glucosidase, from 0.348 mmol PNP in the soil sown with Lh to 0.298 mmol PNP in the soil sown with Pr; and that of arylsulfatase, from 0.164 mmol PNP in the soil sown with Pp to 0.082 mmol PNP in the control soil. The average enzymatic activity of soil sown with fodder grasses was 161% higher than in nonsown soil, and of soil sown with lawn grasses was 83% higher than in nonsown soil. Generally, the enzymatic activity of soils sown with fodder grasses was higher (by 30%) than that of the soils overgrown with the lawn grasses. The lowest enzymatic activity was determined in the control soil not sown with grasses (Figure 10); there was a significantly higher one in the soils sown with Pr, Fr, and Fa; and the highest one was in the soils sown with Lh, Lp, and PP.

Table 5. Enzymatic activity in 1 kg d.m. of soil per 1 h.

Grass Species	Deh μmol TFF	Cat mol O_2	Ure mmol $N\text{-}NH_4$	Pac	Pal	Glu	Aryl
				mmol PNP			
C	1.38 [e]	0.09 [e]	0.30 [b]	0.79 [d]	0.16 [c]	0.30 [b]	0.08 [d]
Lh	10.29 [a]	0.18 [b]	0.47 [ab]	1.20 [ab]	0.40 [a]	0.35 [a]	0.15 [a]
Fa	2.38 [d]	0.14 [c]	0.60 [ab]	1.18 [ab]	0.32 [b]	0.30 [b]	0.14 [ab]
Pp	3.25 [c]	0.19 [ab]	0.88 [a]	1.08 [bc]	0.33 [b]	0.31 [b]	0.16 [a]
Lp	6.87 [b]	0.15 [c]	0.68 [ab]	1.30 [a]	0.35 [ab]	0.34 [a]	0.11 [bc]
Pr	0.65 [f]	0.21 [a]	0.48 [ab]	0.98 [c]	0.17 [c]	0.30 [b]	0.09 [cd]
Fr	1.74 [de]	0.11 [d]	0.63 [ab]	1.27 [a]	0.14 [c]	0.31 [b]	0.11 [cd]
C	1.38 [z]	0.09 [z]	0.30 [y]	0.79 [y]	0.16 [z]	0.30 [y]	0.08 [y]
T1	5.31 [x]	0.17 [x]	0.65 [x]	1.15 [x]	0.35 [x]	0.32 [x]	0.15 [x]
T2	3.09 [y]	0.16 [y]	0.60 [x]	1.18 [x]	0.22 [y]	0.32 [x]	0.10 [y]

Homogeneous groups were calculated separately for each enzyme groups denoted with letters (a–f) were calculated for the species of grass and groups denoted with letters (x–z) were calculated for the type of grass. C—unsown soil; Lh—*Lolium x hybridum* Hausskn; Fa—*Festuca arundinacea*; Pp—*Phleum pratense*, Lp—*Lolium perenne*, Pr—*Poa pratensis*; Fr—*Festuca rubra*.

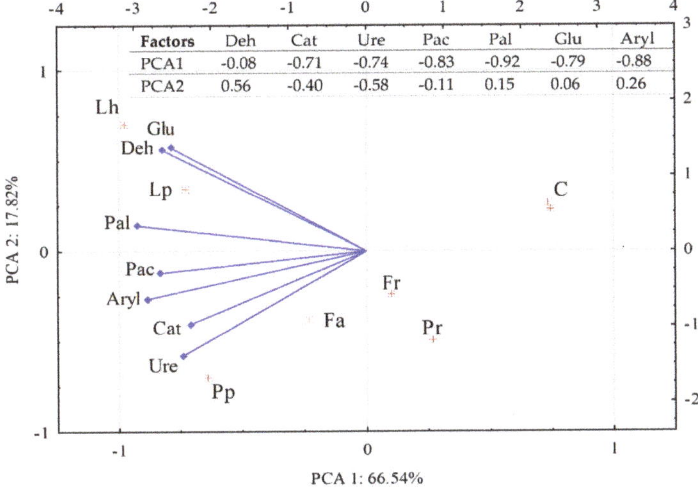

Figure 10. Activity of soil enzymes presented with the principal component analysis (PCA) method. Deh—dehydrogenases; Cat—catalase, Ure—urease; Pac—acid phosphatase; Pal—alkaline phosphatase; Glu—β-glucosidase; Aryl—arylsulfatase. C—unsown soil; Lh—*Lolium x hybridum* Hausskn; Fa—*Festuca arundinacea*; Pp—*Phleum pratense*, Lp—*Lolium perenne*, Pr—*Poa pratensis*; Fr—*Festuca rubra*.

4. Discussion

4.1. Grass Yield

The genotype of grasses turned out to be the main factor which differentiated their yield. Conditions mentioned by Broadbent et al. [69] that model plant growth and development, such as, climatic zone, fraction size, composition of soil, nutrients content in soil, as well as climatic and anthropogenic stresses, could not affect grass growth and development because the pot experiment was performed under controlled conditions. In addition, mineral fertilization was the same for all grass species; therefore this was the genotype that determined the higher biomass produced by the fodder than by the lawn grasses. According to Shukla et al. [70], plants used in agriculture are often grown for green forage and energetic biomass, hence they are increasingly exploited for other purposes than feeds.

4.2. Counts and Diversity of Soil Bacteria

The present study demonstrated that the analyzed grass species modified the soil bacteriobiome to various extents, which was mainly due to the development of their root system [71–73] and chemical composition of their root secretions [13]. According to Berg and Smalla [74] and to Murphy [75], plants may contribute to the establishment of unique communities of soil microorganisms. Among all analyzed grass species, *Lolium* x *hybridum* Hausskn (Lh) contributed to the greatest increase in the population number of organotrophic bacteria compared to the control soil, whereas *Poa pratensis* (Pr), *Lolium perenne* (Lp), *Phleum pratense* (Pp), and *Festuca rubra* (Fr) significantly suppressed their proliferation. The analysis of study results demonstrates that the fodder grasses had a more beneficial effect on the proliferation of organotrophic bacteria and actinobacteria than the lawn grasses had. As reported by Deru et al. [71] and Saleh et al. [73], this could be due to the genetic determinants of individual grass species, which affect development of their root system; whereas the root system influences the development of rhizosphere microbiome by the mineral and organic compounds it secretes [13,74]. Singh et al. [76] emphasized that greater amounts of root secretions produced by young plants contribute to a better availability of carbon and energy sources to microorganisms. In addition, these secretions facilitate rhizosphere colonization by microorganisms [77]. This, in turn, leads to cooperation between the plant and the bacteriobiome, because part of rhizospheric bacteria penetrate inside plant tissues through damaged tissue or due to the release of enzymes capable of increasing solubility of nonabsorbable elements [73].

Although fodder and lawn grasses sown onto the soil elicited changes in the counts of organotrophs and actinobacteria, they did not improve their ecophysiological diversity index (EP). A shift could, however, be noticed in bacteria development towards the k strategists, i.e., slow-growing bacteria, which was indicated by decreased values of the colony development index (CD) caused by both fodder and lawn grasses. These results confirm earlier findings reported by De Leij et al. [61], and Murphy et al. [75], who also observed that the microbiome of the rhizosphere of plants changes along with the prolonging growing season, and that the population of r-strategists turns into k-strategists. Marschner et al. [78], Murphy et al. [75], and Kielak et al. [79] demonstrated that bacterial communities of the rhizosphere are initially predominated by *Proteobacteria* r-strategists. Also in our study, the *Proteobacteria* and *Actinobacteria* were the prevailing phyla on all pots; whereas the prevailing classes included: *Alphaproteobacteria* and *Actinobacteria*; the prevailing orders were: *Actinomycetales, Sphingomycetales,* and *Rhizobiales*; the prevailing families included: *Sphingomonadaceae* and *Hyphomicrobiaceae*; and the predominating genera were: *Kaistobacter, Rhodoplanes, Teracoccus,* and *Flavobacterium*. These results correspond with literature data [22,33,80–83]. In general, a richer bacteriobiome in terms of diversity was demonstrated in the soils sown with grasses than in the control soil without grasses. In the case of the fodder grasses, the greatest diversity occurred in the rhizosphere of *Poa pratensis* (Pr), whereas, in the case of lawn grasses, it was in the rhizosphere of *Phleum pratense* (Pp).

The response of *Proteobacteria* and *Actinobacteria* to sowing grasses onto soil varied. Greater OTUs of *Proteobacteria* were demonstrated in the soils sown with grasses, regardless of their functional type, than in the control soil, whereas the OTU number of *Actinobacteria* in the soil was reduced by both groups of grasses. Changes at the level of phylum and other taxonomic units in the soil sown with various species of legumes and grasses were also observed by Zhou et al. [47] and Singh et al. [76].

A special trait of *Actinobacteria* is their resistance to extreme environmental conditions [22,33,82]. This phylum was described as a promising taxon of plant growth promoters [83].

According to Delgado-Baquerizo et al. [80], the most abundant class of bacteria in soils of the world is *Alphaproteobacteria*, which includes *Bradyrhizobium*, *Sphingomonas*, *Rhodoplanes*, *Devosia*, and *Kaistobacter* genera; whereas among *Actinobacteria*, there are the *Streptomyces*, *Salinibacterium*, and *Mycobacterium* genera. In our study also, the *Kaistobacter* and *Rhodoplanes* genera were found to prevail, but other major genera included *Terracoccus*, *Candidatus Koribacter*, and *Devosia*.

Both the results of this study and literature data [21,22,33,80,82] indicate that investigations addressing the genetic biodiversity of bacteria should be continued in various soil ecosystems.

4.3. Activity of Soil Enzymes

Being sensitive indicators of soil quality, enzymes are strongly associated with the microbiological activity and species colonization of plants [4,84]. In the present study, grasses stimulated the biochemical activity of soil. This is due to their beneficial effect on the soil bacteriobiome, as indicated by results of this study and by literature data [51,52,84–86]. The association between the activity of soil enzymes and microbiome quality is due to the origin of enzymes [38,50,53,84]. In soil ecosystems, they are mainly derived from microorganisms and, to a lesser extent, from plants and other soil organisms [87–90]. The positive correlation between the activity of soil enzymes and the activity of microorganisms has been demonstrated by many experts in soil science [91–93]. In our own research, the higher activity of soil enzymes in soil sown with fodder grass is mainly associated with a greater diversity of bacteria at the family and genus level in soil from below these plants than in soil from below lawn grasses. The values of the Shannon-Wiener and Simpson indicators prove this. Nevertheless, the more beneficial effect elicited by the fodder than by the lawn grasses on the biochemical properties of soil proves that, by activating the microbiome, the plants can intermediately affect enzymatic activity. This hypothesis was corroborated by other authors [3,86,91–94].

5. Conclusions

The analyzed grass species from the family *Poacea* had a beneficial effect on soil microbiome and activity of soil enzymes. The intensity of their effect was determined by both their species and their functional type. More favorable conditions for the growth and development of soil bacteria, and thereby for the enhanced enzymatic activity, were offered by the fodder than by the lawn grasses. Among the fodder grasses, the greatest bacteriobiome diversity was demonstrated in the soil sown with *Poa pratensis* (Pp), whereas, among the lawn grasses, it was in soil sown with *Phleum pretense* (Pr). The highest enzymatic activity was determined also. Considering the fodder grasses, this was in the soil with *Lolium x hybridum* Hausskn (Lh), and in the soil with *Lolium perenne* (Lp) in the case of lawn grasses. The sowing of soils with grasses caused the succession of bacterial communities from r strategy to k strategy. In all pots, the prevailing phyla included *Proteobacteria* and *Actinobacteria*; the prevailing classes were *Alphaproteobacteria* and *Actinobacteria*; the prevailing orders included *Actinomycetales*, *Sphingomycetales*, and *Rhizobiales*; the prevailing families were *Sphingomonadaceae* and *Hyphomicrobiaceae*; and the prevailing genera included *Kaistobacter*, *Rhodoplanes*, *Teracoccus*, and *Flavobacterium*.

Supplementary Materials: The following are available online at http://www.mdpi.com/1424-2818/12/6/212/s1, Table S1: One-way significance tests carried out using the analysis of variance (ANOVA), Figure S1: The relative abundance of dominant phylum bacteria in soil. Data on the number of readings greater than 1% of all OTUs, Figure S2: The relative abundance of dominant class bacteria in soil. Data on the number of readings greater than 1% of all OTUs, Figure S3: The relative abundance of dominant order bacteria in soil. Data on the number of readings greater than 1% of all OTUs, Figure S4: The relative abundance of dominant family bacteria in soil. Data on the number of readings greater than 1% of all OTUs.

Author Contributions: A.B. conceived and designed the ideas and. wrote the manuscript with the help of J.W. and J.K.; A.B. conducted the experiments, collected and analyzed the data, conducted the bioinformatic analysis and visualization of data; all authors contributed to the final version of this manuscript. All authors have read and agreed to the published version of the manuscript.

Funding: This study was supported by the Ministry of Science and Higher Education funds for statutory activity. Project financially supported by Minister of Science of Higher Education in the range of the program entitled "Regional Initiative of Excellence" for the years 2019–2022, Project No. 010/RID/2018/19, amount of funding 12,000,000 PLN.

Conflicts of Interest: The authors declare no conflict of interest. The funders had no role in the design of the study; in the collection, analyses, or interpretation of data; in the writing of the manuscript, or in the decision to publish the results.

References

1. Berendsen, R.L.; Pieterse, C.M.J.; Bakker, P.A.H.M. The rhizosphere microbiome and plant health. *Trends Plant Sci.* **2012**, *17*, 478–486. [CrossRef] [PubMed]
2. Naylor, D.; Coleman-Derr, D. Drought Stress and Root-Associated Bacterial Communities. *Front. Plant Sci.* **2018**, *8*, 2223. [CrossRef] [PubMed]
3. Xavier, C.V.; Moitinho, M.R.; De Bortoli Teixeira, D.; André de Araújo Santos, G.; de Andrade Barbosa, M.; Bastos Pereira Milori, D.M.; Rigobelo, E.; Corá, J.E.; La Scala Júnior, N. Crop rotation and succession in a no-tillage system: Implications for CO_2 emission and soil attributes. *J. Environ. Manag.* **2019**, *1*, 8–15. [CrossRef] [PubMed]
4. Vives-Peris, V.; de Ollas, C.; Gómez-Cadenas, A.; Pérez-Clemente, R.M. Root exudates: From plant to rhizosphere and beyond. *Plant Cell Rep.* **2019**, *25*. [CrossRef]
5. Evans, S.E.; Wallenstein, M.D. Climate change alters ecological strategies of soil bacteria. *Ecol. Lett.* **2013**, *17*, 155–164. [CrossRef]
6. Fierer, N. Embracing the unknown: Disentangling the complexities of the soil microbiome. *Nat. Rev. Microbiol.* **2017**, *15*, 579–590. [CrossRef]
7. Singh, B.K.; Bardgett, R.D.; Smith, P.; Reay, D.S. Microorganisms and climate change: Terrestrial feedbacks and mitigation options. *Nat. Rev. Microbiol.* **2010**, *8*, 779–790. [CrossRef]
8. Nannipieri, P.; Ascher, J.; Ceccherini, M.T.; Landi, L.; Pietramellara, G.; Renella, G.; Valori, F. Effects of Root Exudates in Microbial Diversity and Activity in Rhizosphere Soils. In *Molecular Mechanisms of Plant and Microbe Coexistence*; Nautiyal, C.S., Dion, P., Eds.; Springer: Berlin/Heidelberg, Germany, 2008; Volume 15, pp. 339–365. [CrossRef]
9. Candan, N.; Cakmak, I.; Ozturk, L. Zinc-biofortified seeds improved seedling growth under zinc deficiency and drought stress in durum wheat. *J. Plant Nutr. Soil Sci.* **2018**, *181*, 388–395. [CrossRef]
10. Dubey, R.K.; Tripathi, V.; Prabha, R.; Chaurasia, R.; Singh, D.P.; Rao, C.S.; El-Keblawy, A.; Abhilash, P.C. Belowground Microbial Communities: Key Players for Soil and Environmental Sustainability. In *Unravelling the Soil Microbiome. Springer Briefs in Environmental Science*; Springer: Cham, Switzerland, 2020; Volume 2, pp. 5–22. [CrossRef]
11. Schloter, M.; Nannipieri, P.; Sørensen, S.J.; van Elsas, J.D. Microbial indicators for soil quality. *Biol. Fertil. Soils* **2018**, *54*, 1–10. [CrossRef]
12. Lau, J.A.; Lennon, J.T. Rapid responses of soil microorganisms improve plant fitness in novel environments. *Proc. Natl. Acad. Sci. USA* **2012**, *109*, 14058–14062. [CrossRef]
13. Walker, T.S. Root Exudation and Rhizosphere Biology. *Plant Physiol.* **2003**, *132*, 44–51. [CrossRef] [PubMed]
14. Wu, B.; Wang, Z.; Zhao, Y.; Gu, Y.; Wang, Y.; Yu, J.; Xu, H. The performance of biochar-microbe multiple biochemical material on bioremediation and soil micro-ecology in the cadmium aged soil. *Sci. Total Environ.* **2019**, *686*, 719–728. [CrossRef] [PubMed]
15. Hammerbacher, A.; Coutinho, T.A.; Gershenzon, J. Roles of plant volatiles in defence against microbial pathogens and microbial exploitation of volatiles. *Plant Cell Environ.* **2019**, 1–17. [CrossRef] [PubMed]
16. Van de Wouw, A.P.; Idnurm, A. Biotechnological potential of engineering pathogen effector proteins for use in plant disease management. *Biotechnol. Adv.* **2019**, *37*, 107387. [CrossRef] [PubMed]
17. Ponge, J.F. Humus forms in terrestrial ecosystems: A framework to biodiversity. *Soil Biol. Biochem.* **2003**, *35*, 935–945. [CrossRef]
18. Ponge, J.F. Plant–soil feedbacks mediated by humus forms: A review. *Soil Biol. Biochem.* **2013**, *57*, 1048–1060. [CrossRef]

19. Rousk, J.E.; Bääth, P.C.; Brookes, C.L.; Lauber, C.; Lozupone, J.G.; Caporaso, R.; Knight, R.; Fierer, N. Soil bacterial and fungal communities across a pH gradient in an arable soil. *ISME J.* **2010**, *4*, 10, 1340–1351. [CrossRef]

20. Rillig, M.; Muller, L.; Lehmann, A. Soil aggregates as massively concurrent evolutionary incubators. *ISME J.* **2017**, *11*, 1943–1948. [CrossRef]

21. Shi, Y.; Li, Y.T.; Xiang, X.J.; Sun, R.B.; Yang, T.; He, D.; Zhang, K.P.; Ni, Y.Y.; Zhu, Y.G.; Adams, J.M.; et al. Spatial scale affects the relative role of stochasticity versus determinism in soil bacterial communities in wheat fields across the North China Plain. *Microbiome* **2018**, *6*, 27. [CrossRef]

22. Shi, Y.; Li, Y.; Yuan, M.; Adams, J.M.; Pan, X.; Yang, Y.; Chu, H. A biogeographic map of soil bacterial communities in wheats field of the North China Plain. *Soil Ecol. Lett.* **2019**, *1*, 50–58. [CrossRef]

23. Barrios, E. Soil biota, ecosystem services and land productivity. *Ecol. Econom.* **2007**, *64*, 269–285. [CrossRef]

24. Nannipieri, P.; Ascher, J.; Ceccherini, M.T.; Landi, L.; Pietramellara, G.; Renella, G. Microbial diversity and soil functions. *Eur. J. Soil Sci.* **2017**, *68*, 12–26. [CrossRef]

25. Borowik, A.; Wyszkowska, J. Soil moisture as a factor affecting the microbiological and biochemical activity of soil. *Plant Soil Environ.* **2016**, *62*, 250–255. [CrossRef]

26. Borowik, A.; Wyszkowska, J. Impact of temperature on the biological properties of soil. *Int. Agrophys.* **2016**, *30*, 1–8. [CrossRef]

27. Nannipieri, P.; Trasar-Cepeda, C.; Dick, R.P. Soil enzyme activity: A brief history and biochemistry as a basis for appropriate interpretations and meta-analysis. *Biol. Fertil. Soils* **2017**, *54*, 11–19. [CrossRef]

28. VCU 2019, Value for Cultivation and Use (VCU) BSPB Plant Breeding Matters. Available online: http://plantbreedingmatters.com/evaluation.php (accessed on 3 February 2020).

29. Raman, J.K.; Alves, C.M.; Gnansounou, E. A review on moringa tree and vetiver grass—Potential biorefinery feedstocks. *Bioresour. Technol.* **2018**, *249*, 1044–1051. [CrossRef]

30. Peeters, A. Importance, evolution, environmental impact and future challenges of grasslands and grassland-based systems in Europe. *Grassl. Sci.* **2009**, *55*, 113–125. [CrossRef]

31. EEA (European Environment Agency). 2019. Available online: https://www.eea.europa.eu/data-and-maps/indicators/progress-in-management-of-contaminated-sites-3 (accessed on 20 February 2020).

32. European Commission. 2019. Available online: www.ec.europa.eu/environment/soil/index_en.htm (accessed on 20 February 2020).

33. Zhang, X.; Xu, S.; Li, C.; Zhao, L.; Feng, H.; Yue, G.; Ren, Z.; Cheng, G. The soil carbon/nitrogen ratio and moisture affect microbial community structures in alkaline permafrost-affected soils with different vegetation types on the Tibetan plateau. *Res. Microbiol.* **2014**, *165*, 128–139. [CrossRef]

34. Borowik, A.; Wyszkowska, J. Bioaugmentation of soil contaminated with diesel oil. *J. Elem.* **2018**, *23*, 1161–1178. [CrossRef]

35. Chikere, C.B.; Okpokwasili, G.C.; Chikere, B.O. Monitoring of microbial hydrocarbon remediation in the soil. *3 Biotech.* **2011**, *1*, 3. [CrossRef]

36. Lipińska, A.; Wyszkowska, J.; Kucharski, J. Diversity of organotrophic bacteria, activity of dehydrogenases and urease as well as seed germination and root growth *Lepidium sativum, Sorghum saccharatum* and *Sinapis alba* under the influence of polycyclic aromatic hydrocarbons. *Environ. Sci. Pollut. Res. Int.* **2015**, *22*, 18519–18530. [CrossRef] [PubMed]

37. Telesiński, A.; Krzyśko-Łupicka, T.; Cybulska, K.; Wróbel, J. Response of soil phosphatase activities to contamination with two types of tar oil. *Environ. Sci. Pollut. Res.* **2018**, *25*, 28642–28653. [CrossRef] [PubMed]

38. Xu, X.; Liu, W.; Tian, S.; Wang, W.; Qi, Q.; Jiang, P.; Gao, X.; Li, F.; Yu, H. Petroleum hydrocarbon-degrading bacteria for the remediation of oil pollution under aerobic conditions: A perspective analysis. *Front. Microbiol.* **2018**, *9*, 2885. [CrossRef] [PubMed]

39. Zaborowska, M.; Kucharski, J.; Wyszkowska, J. Biochemical and microbiological activity of soil contaminated with o-cresol and biostimulated with *Perna canaliculus* mussel meal. *Environ. Monit Assess.* **2018**, *190*, 602. [CrossRef] [PubMed]

40. Kucharski, J.; Wieczorek, K.; Wyszkowska, J. Changes in the enzymatic activity in sandy loam soil exposed to zinc pressure. *J. Elem.* **2011**, *16*, 577–589. [CrossRef]

41. Wyszkowska, J.; Boros-Lajszner, E.; Borowik, A.; Baćmaga, M.; Kucharski, J.; Tomkiel, M. Implication of zinc excess on soil health. *J. Environ. Sci. Health B* **2016**, *51*, 261–270. [CrossRef]

42. Zaborowska, M.; Kucharski, J.; Wyszkowska, J. Biological activity of soil contaminated with cobalt, tin and molybdenum. *Environ. Monit. Assess.* **2016**, *188*, 398. [CrossRef]

43. Huang, Y.; Xiao, L.; Li, F.; Xiao, M.; Lin, D.; Long, X.; Wu, Z. Microbial degradation of pesticide residues and an emphasis on the degradation of cypermethrin and 3-phenoxy benzoic acid: A review. *Molecules* **2018**, *23*, 2313. [CrossRef]

44. Niewiadomska, A.; Sulewska, H.; Wolna-Maruwka, A.; Waraczewska, Z.; Budka, A.; Ratajczak, K. An assessment of the influence of selected herbicides on the microbial parameters of soil in maize (*Zea mays*) cultivation. *Appl. Ecol. Env. Res.* **2018**, *16*, 4735–4752. [CrossRef]

45. Tomkiel, M.; Baćmaga, M.; Borowik, A.; Kucharski, J.; Wyszkowska, J. Effect of a mixture of flufenacet and isoxaflutole on population numbers of soil-dwelling microorganisms, enzymatic activity of soil, and maize yield. *J. Environ. Sci. Health B* **2019**, *1*, 11. [CrossRef]

46. Borowik, A.; Wyszkowska, J. Response of *Avena sativa* L. and the soil microbiota to the contamination of soil with shell diesel oil. *Plant Soil Environ.* **2018**, *64*, 102–107. [CrossRef]

47. Zhou, Y.; Qin, Y.; Liu, X.; Feng, Z.; Zhu, H.; Yao, Q. Soil bacterial function associated with stylo (Legume) and bahiagrass (grass) is affected more strongly by soil chemical property than by bacterial community composition. *Front. Microbiol.* **2019**, *10*, 798. [CrossRef] [PubMed]

48. Borowik, A.; Wyszkowska, J.; Oszust, K. Functional diversity of fungal communities in soil contaminated with diesel oil. *Front. Microbiol.* **2017**, *8*, 1862. [CrossRef]

49. Tshikantwa, T.S.; Ullah, M.W.; He, F.; Yang, G. Current trends and potential applications of microbial interactions for human welfare. *Front. Microbiol.* **2018**, *9*, 1156. [CrossRef] [PubMed]

50. Bell, C.W.; Fricks, B.E.; Rocca, J.D.; Steinweg, J.M.; McMahon, S.K.; Wallenstein, M.D. High-throughput fluorometric measurement of potential soil extracellular enzyme activities. *J. Vis. Exp.* **2013**, e50961. [CrossRef] [PubMed]

51. Moeskops, B.; Buchan, D.; Sleutel, S.; Herawaty, L.; Husen, E.; Saraswati, R.; Setyorini, D.; De Neve, S. Soil microbial communities and activities under intensive organicand conventional vegetable farming in west java, Indonesia. *Appl. Soil Ecol.* **2010**, *45*, 112–120. [CrossRef]

52. Zhan, X.; Wu, W.; Zhou, L.; Liang, J.; Jiang, T. Interactive effect of dissolved organic matter and phenanthrene on soil enzymatic activities. *J. Environ. Sci.* **2010**, *22*, 607–614. [CrossRef]

53. Knight, T.R.; Dick, R.P. Differentiating microbial and stabilized β-glucosidase activity relative to soil quality. *Soil Biol. Biochem.* **2004**, *36*, 2089–2096. [CrossRef]

54. Wyszkowska, J.; Wyszkowski, M. Activity of soil dehydrogenases, urease and acid and alkaline phosphatase in soil polluted with petroleum. *J. Toxicol. Environ. Health Part A* **2010**, *73*, 1202–1210. [CrossRef]

55. Vong, P.C.; Piutti, S.; Slezack-Deschaumes, S.; Beniziri, E.; Guckert, A. Sulphur immobilization and arylsulphatase activity in two calcareous arable and fallow soils as affected by glucose additions. *Geoderma* **2008**, *148*, 79–84. [CrossRef]

56. World Reference Base for Soil Resources. *International Soil Classification System for Naming Soils and Creating Legends for Soil Maps*; World Soil Resources Reports No. 106; FAO: Rome, Italy, 2014.

57. Borowik, A.; Wyszkowska, J.; Wyszkowski, M. Resistance of aerobic microorganisms and soil enzyme response to soil contamination with Ekodiesel Ultra fuel. *Environ. Sci. Pollut. Res.* **2017**, *24*, 24346–24363. [CrossRef]

58. Wyszkowska, J.; Borowik, A.; Olszewski, J.; Kucharski, J. Soil bacterial community and soil enzyme activity depending on the cultivation of *Triticum aestivum, Brassica napus,* and *Pisum sativum* ssp. *arvense. Diversity* **2019**, *11*, 246. [CrossRef]

59. Bunt, J.S.; Rovira, A.D. Microbiological studies of some subantarctic soils. *J. Soil Sci.* **1955**, *6*, 119–128. [CrossRef]

60. Parkinson, D.; Gray, F.R.G.; Williams, S.T. *Methods for Studying the Ecology of Soil Microorganism*; IBP Handbook 19; Blackwell Scientific Publication: Oxford, UK, 1971.

61. De Leij, F.A.A.M.; Whipps, J.M.; Lynch, J.M. The use of colony development for the charactization of bacterial communities in soil and on roots. *Microb. Ecol.* **1994**, *27*, 81–97. [CrossRef] [PubMed]

62. Öhlinger, R. Dehydrogenase Activity with the Substrate TTC. In *Methods in Soil Biology*; Schinner, F., Ohlinger, R., Kandler, E., Margesin, R., Eds.; Springer: Berlin, Germany, 1996; pp. 241–243.

63. Johnson, J.I.; Temple, K.L. Some variables affecting the measurement of catalase activity in soil. *Soil Sci. Soc. Am. J.* **1964**, *28*, 207–216. [CrossRef]

64. Alef, K.; Nannipieri, P. *Methods in Applied Soil Microbiology and Biochemistry*; Alef, K., Nannipieri, P., Eds.; Academic London: London, UK, 1988; pp. 316–365.

65. De Santis, T.Z.; Hugenholtz, P.; Larsen, N.; Rojas, M.; Brodie, E.L.; Keller, K.; Huber, T.; Dalevi, D.; Hu, P.; Andersen, G.L. Greengenes, a chimera-checked 16S rRNA gene database and workbench compatible with ARB. *Appl. Environ. Microbiol.* **2006**, *72*, 5069–5072. [CrossRef]

66. Dell Inc. *Dell Statistica (Data Analysis Software System), Version 13.1*; Dell Inc.: Tulsa, OK, USA, 2016.

67. Parks, D.H.; Tyson, G.W.; Hugenholtz, P.; Beiko, R.G. STAMP: Statistical analysis of taxonomic and functional profiles. *Bioinformatics* **2014**, *30*, 3123–3124. [CrossRef]

68. Krzywinski, M.I.; Schein, J.E.; Birol, I.; Connors, J.; Gascoyne, R.; Horsman, D.; Jones, S.J.; Marra, M.A. Circos: An information aesthetic for comparative genomics. *Genome Res.* **2009**, *19*, 1639–1645. [CrossRef]

69. Broadbent, A.A.D.; Stevens, C.J.; Ostle, N.J.; Orwin, K.H. Biogeographic differences in soil biota promote invasive grass response to nutrient addition relative to co-occurring species despite lack of belowground enemy release. *Oecologia* **2018**, *186*, 611–620. [CrossRef]

70. Shukla, P.; Chaurasia, P.; Younis, K.; Qadri, O.S.; Faridi, S.A.; Srivastava, G. Nanotechnology in sustainable agriculture: Studies from seed priming to post-harvest management. *Nanotechnol. Environ. Eng.* **2019**, *4*, 1. [CrossRef]

71. Deru, J.; Schilder, H.; Van der Schoot, J.R.; Van Eekeren, N. No Trade-off between Root Biomass and Aboveground Production in Lolium perenne. In *Breeding in a World of Scarcity*; Roldán-Ruiz, I., Baert, J., Reheul, D., Eds.; Springer: Cham, Switzerland, 2016; Volume 43, pp. 289–292. [CrossRef]

72. Raaijmakers, J.M.; Paulitz, C.T.; Steinberg, C.; Alabouvette, C.; Moenne-Loccoz, Y. The rhizosphere: A playground and battlefield for soilborne pathogens and beneficial microorganisms. *Plant Soil* **2009**, *321*, 341–361. [CrossRef]

73. Saleh, D.; Jarry, J.; Rani, M.; Aliferis, K.; Seguin, P.; Jabaji, S.H. Diversity, distribution and multi-functional attributes of bacterial communities associated with the rhizosphere and endosphere of timothy *(Phleum pratense L.). J. Appl. Microbiol.* **2019**, *127*, 794–811. [CrossRef] [PubMed]

74. Berg, G.; Smalla, K. Plant species and soil type cooperatively shape the structure and function of microbial communities in the rhizosphere. *FEMS Microbiol. Ecol.* **2009**, *68*, 1–13. [CrossRef] [PubMed]

75. Murphy, C.A.; Foster, B.L.; Gao, C. temporal dynamics in rhizosphere bacterial communities of three perennial grassland species. *Agronomy* **2016**, *6*, 17. [CrossRef]

76. Singh, B.K.; Munro, S.; Potts, J.M.; Millard, P. Influence of grass species and soil type on rhizosphere microbial structure in grassland soils. *Appl. Soil Ecol.* **2007**, *36*, 147–155. [CrossRef]

77. Chen, K.; Chang, Y.; Chiou, W. Remediation of diesel-contaminated soil using in situ chemical oxidation (ISCO) and the effects of common oxidants on the indigenous microbial community: A comparison study. *J. Chem. Technol. Biotechnol.* **2016**, *91*, 1877–1888. [CrossRef]

78. Marschner, P.; Neumann, G.; Kania, A.; Weiskopf, L.; Lieberei, R. Spatial and temporal dynamics of the microbial community structure in the rhizosphere of cluster roots of white lupin (*Lupinus albus* L.). *Plant Soil* **2002**, *246*, 167–174. [CrossRef]

79. Kielak, A.; Pijl, A.S.; van Veen, J.A.; Kowalchuk, G.A. Differences in vegetation composition and plant species identity lead to only minor changes in soil-borne microbial communities in a former arable field. *FEMS Microbiol. Ecol.* **2008**, *63*, 372–382. [CrossRef]

80. Delgado-Baquerizo, M.; Oliverio, A.M.; Brewer, T.E.; Benavent-González, A.; Eldridge, D.J.; Bardgett, R.D.; Maestre, F.T.; Singh, B.K.; Fierer, N. A global atlas of the dominant bacteria found in soil. *Science* **2018**, *359*, 320–325. [CrossRef]

81. Ghuneim, L.-A.J.; Jones, D.L.; Golyshin, P.N.; Golyshina, O.V. Nano-sized and filterable bacteria and archaea: Biodiversity and function. *Front. Microbiol.* **2018**, *9*, 1971. [CrossRef]

82. Chu, H.; Sun, H.; Tripathi, B.M.; Adams, J.M.; Huang, R.; Zhang, Y.; Shi, Y. Bacterial community dissimilarity between the surface and subsurface soils equals horizontal differences over several kilometers in the western Tibetan Plateau. *Environ. Microbiol.* **2016**, *18*, 1523–1533. [CrossRef] [PubMed]

83. Hamedi, J.; Mohammadipanah, F. Biotechnological application and taxonomical distribution of plant growth promoting actinobacteria. *J. Ind. Microbiol. Biotechnol.* **2015**, *42*, 157–171. [CrossRef] [PubMed]

84. Alkorta, I.; Aizpurua, A.; Riga, P.; Albizu, I.; Amézaga, I.; Garbisu, C. Soil enzyme activities as biological indicators of soil health. *Rev. Environ. Health* **2003**, *18*, 1. [CrossRef] [PubMed]

85. Das, N.; Chandran, P. Microbial degradation of petroleum hydrocarbon contaminants: An overview. *Biotechnol. Res. Int.* **2011**, *2011*, 941810. [CrossRef]

86. Borowik, A.; Wyszkowska, J.; Kucharski, M.; Kucharski, J. Implications of soil pollution with diesel oil and BP petroleum with ACTIVE Technology for soil health. *Int. J. Environ. Res. Public Health* **2019**, *16*, 2474. [CrossRef] [PubMed]

87. Caldwell, B.A. Enzyme activities as a component of soil biodiversity: A review. *Pedobiologia* **2005**, *49*, 637–644. [CrossRef]

88. Niewiadomska, A.; Sulewska, H.; Wolna-Maruwka, A.; Ratajczak, K.; Waraczewska, Z.; Budka, A.; Głuchowska, K. The influence of biostimulants and foliar fertilisers on the process of biological nitrogen fixation and the level of soil biochemical activity in soybean (Glycine Maxl.) cultivation. *Appl. Ecol. Environ. Res.* **2019**, *17*, 12649–12666. [CrossRef]

89. Oleszczuk, P.; Jośko, I.; Futa, B.; Pasieczna-Patkowska, S.; Pałys, E.; Kraska, P. Effect of pesticides on microorganisms, enzymatic activity and plant in biochar-amended soil. *Geoderma* **2014**, *214—215*, 10–18. [CrossRef]

90. Zhao, F.Z.; Ren, C.J.; Han, X.H.; Yang, G.H.; Wang, J.; Doughty, R. Changes of soil microbial and enzyme activities are linked to soil C, N and P stoichiometry in afforested ecosystems. *For. Ecol. Manag.* **2018**, *427*, 289–295. [CrossRef]

91. Wang, X.; Song, D.; Liang, G.; Zhang, Q.; Ai, C.; Zhou, W. Maize biochar addition rate influences soil enzyme activity and microbial community composition in a fluvo-aquic soil. *Appl. Soil Ecol.* **2015**, *96*, 265–272. [CrossRef]

92. Li, J.; Zhou, X.; Yan, J.; Li, H.; He, J. Effects of regenerating vegetation on soil enzyme activity and microbial structure in reclaimed soils on a surface coal mine site. *Appl. Soil Ecol.* **2015**, *87*, 56–62. [CrossRef]

93. Raiesi, F.; Beheshti, A. Microbiological indicators of soil quality and degradation following conversion of native forests to continuous croplands. *Ecol. Indic.* **2015**, *50*, 173–185. [CrossRef]

94. García-Gaytán, V.; Hernández-Mendoza, F.; Coria-Téllez, A.; García-Morales, S.; Sánchez-Rodríguez, E.; Rojas-Abarca, L.; Daneshvar, H. Fertigation: Nutrition, Stimulation and Bioprotection of the Root in High Performance. *Plants* **2018**, *7*, 88. [CrossRef] [PubMed]

Article

Soil Bacterial Community and Soil Enzyme Activity Depending on the Cultivation of *Triticum aestivum*, *Brassica napus*, and *Pisum sativum* ssp. *arvense*

Jadwiga Wyszkowska [1,*], Agata Borowik [1], Jacek Olszewski [2] and Jan Kucharski [1]

[1] Department of Microbiology, University of Warmia and Mazury in Olsztyn, 10-727 Olsztyn, Poland; agata.borowik@uwm.edu.pl (A.B.); jan.kucharski@uwm.edu.pl (J.K.)
[2] Didactic and Experimental Centre, University of Warmia and Mazury in Olsztyn, 10-727 Olsztyn, Poland; jacek.olszewski@uwm.edu.pl
* Correspondence: jadwiga.wyszkowska@uwm.edu.pl

Received: 19 November 2019; Accepted: 15 December 2019; Published: 17 December 2019

Abstract: This study aims to determine the effects of crops and their cultivation regimes on changes in the soil microbiome. Three plant species were selected for the study: *Triticum aestivum*, *Brassica napus*, and *Pisum sativum* ssp. *arvense*, that were cultivated in soils with a similar particle size fraction. Field experiments were performed on the area of the Iławski Lake District (north-eastern Poland) at the Production and Experimental Station 'Bałcyny' (53°35′49″ N, 19°51′20″ E). In soil samples counts, organotrophic bacteria and actinobacteria were quantified, and the colony development index (CD) and ecophysiological diversity index (EP) were computed. In addition, a 16S amplicon sequencing encoding gene was conducted based on the hypervariable region V3–V4. Further analyses included an evaluation of the basic physiochemical properties of the soil and the activities of dehydrogenases, catalase, urease, acid phosphatase, alkaline phosphatase, arylsulfatase, and β-glucosidase. Analyses carried out in the study demonstrated that the rhizosphere of *Triticum aestivum* had a more beneficial effect on bacteria development than those of *Brassica napus* and *Pisum sativum* ssp. *arvense*, as indicated by the values of the ecophysiological diversity index (EP) and operational taxonomic unit (OTU) abundance calculated for individual taxa in the soils in which the studied crops were grown. More OTUs of the taxa *Alphaproteobacteria*, *Gammaproteobacteria*, *Clostridia*, *Sphingomonadales*, *Rhodospirillales*, *Xanthomonadales*, *Streptomycetaceae*, *Pseudonocardiaceae*, *Acetobacteraceae*, *Solibacteraceae*, *Kaistobacter*, *Cohnella*, *Azospirillum*, *Cryptosporangium*, *Rhodoplanes*, and *Saccharopolyspora* were determined in the bacteriome structure of the soil from *Triticum aestivum* cultivation than in the soils from the cultivation of *Brassica napus* and *Pisum sativum* ssp. *arvense*. Also, the activities of most of the analyzed enzymes, including urease, catalase, alkaline phosphatase, β-glucosidase, and arylsulfatase, were the higher in the soil sown with *Triticum aestivum* than in those with the other two plant species.

Keywords: bacteria; diversity; operational taxonomic unit (OTU); enzymes activity

---

1. Introduction

Growing attention is being paid to the proper functioning of ecosystems [1,2]. This is due to numerous initiatives which have been undertaken globally to make the population aware of the need to protect soils and for their sustainable management. One of the first of these initiatives was the announcement of the European Soil Health Card, which has made both the population and policy makers aware of the significant role of the soil environment, and has introduced a new term, 'environmental services', which was later changed to 'ecosystem services' [3]. The concept of soil quality and health, often used interchangeably, is becoming increasingly recognized worldwide [4–6]. Legaz et al. [6] and Bünemann et al. [7] define 'soil quality' as the capability of soil to function in the framework of ecosystems, and also as land management and promotion of the biodiversity and health of plants and animals. According to Cardoso et al. [5] and Veum et al. [8], the term 'soil health' highlights the fact that soil is a live and dynamic being whose functions are determined by the diversity of living organisms. Therefore, the physical, chemical, and biological properties of soil may change according to biotic and abiotic factors, which consequently affect its functions and ecosystemic services [3,9].

Apart from elements such as profile morphology, physical and chemical properties, and microclimate, the microbiological activity of soil is another principal factor affecting its fertility [10]. Microorganisms and enzymes determine the transformation of organic matter introduced into soil, influence the humification process, and play a key role in the biogeochemical cycles of many macro- and micro- elements [7]. They are key participants in most of the important cycles, including carbon, nitrogen, phosphorus, and sulfur circulation. It is mainly the biogeochemical cycles of these elements that determines the quality of the natural environment, including the soil environment [9]. The role of microbes is invaluable in the transformation of postharvest residues, as well as natural and organic fertilizers [11,12], the detoxication of organic contaminants in soil [13,14], minimizing the prevalence of pests and pathogens [11], and finally, in establishing symbiotic systems with plants [15]. The enormous importance of microorganisms for the soil environment and their high metabolic activity are confirmed by the fact that the microbial biomass of soil accounts for approximately 85% of the total biomass of all living organisms colonizing this environment [11].

The main factors which affect the development of soil microorganisms include the abundance of organic and mineral colloids in the soil, the climate and microclimate, oxidation associated with humidity status [16–19], soil pH [20,21], soil tillage and plant cultivation systems [22–24], and fertilization systems [25]. An inseparable element in this case is soil temperature, which is affected by the climate and microclimate [16,26,27]. Significant are also contamination with heavy metals [28,29], various hydrocarbons [30–32], plant protection agents [33–35], and dioxins, as well as the salinity level [36], all of which determine the proliferation of various microbial communities for which an increase their diversity may reduce soil fertility and, ultimately, influence soil productivity [37].

All the aforementioned parameters determine the development of not only microorganisms, but plants as well. The connection between both of these groups of organisms is of the utmost importance, e.g., in the case of symbioses of certain bacteria and fungi species with plants. Noteworthy are also the microorganisms that live in association with plants and produce growth hormones, bind atmospheric nitrogen, and protect plants against pathogens. An important role is also ascribed to the microbes classified as plant growth-promoting rhizobacteria (PGPR) [38]. On the other hand, apart from bacteriorrhiza, mycrorrhiza, and microbe associations with plants, we cannot observe a beneficial effect of root secretions on microorganisms or a positive impact of root systems on the physical properties of soil, as it serves the function of drainage, which is essential for appropriate soil oxygenation [27,39]. Root secretions, which may contain organic acids, amino acids, carbohydrates, vitamins, and metal ions, modify the microenvironments of the rhizosphere [40]. They release ions, oxygen, water, and carbon-containing compounds [41]. They may both stimulate and inhibit the development of a soil microbiome [42], and act both as repellents [43] and attractants [44]. Volatile organic compounds (VOCs) emitted by soil microorganisms can affect root growth. Bacterial volatile compounds (BVCs) are used as a source of nutrients and information in plant–bacterium interactions [15]. Such interactions are the strongest in the rhizosphere [7,45], which is colonized by 10 to 100 times more microorganisms than the sphere that is distant from plant roots [46]. The beneficial bacteria of the rhizosphere compete with other microorganisms for organic compounds and colonize plant roots [47]. Microorganisms colonizing this microecosystem are affected by the root secretions of particular plant species and form specific microbial communities [48]. The 16S amplicon sequencing analysis, a small subunit ribosomal ribonucleic acid (SSU rRNA), is an important element in biological quality assessments of soil, as it may provide answers about the response of individual taxa to variable factors in different agricultural ecosystems [49]. Despite the severity of the problem, investigations into the effects of plants on soil bacteria based on next-generation methods are still scarce.

The complex nature of the active rhizosphere has become a premise for undertaking a study aimed at determining the effect of a crop and its cultivation regime on changes in the soil microbiome. Three plant species were used in the study: winter wheat, winter rape, and field pea; these species were cultivated on soil with a similar fraction size. Analyses were performed to determine the structure and diversity of microorganisms and the activity of the soil enzymes participating in the metabolism of carbon, nitrogen, phosphorus, and sulfur. The coupled use of microbiological, biochemical, and physicochemical parameters form the basis for a better understanding of the interactions between the rhizosphere, microorganisms, and plants, which is extremely important from the viewpoint of soil fertility.

2. Materials and Methods

2.1. Sampling Area

Field experiments were performed in the area of the Iławski Lake District (north-eastern Poland) at the Production and Experimental Station 'Bałcyny' (53°35′49″ N, 19°51′20″ E) of the University of Warmia and Mazury in Olsztyn (Poland). The Lake District stretches over 4230 km^2 and is one of the least polluted regions of Poland, called the Green Lungs of Poland. It is characterized by young glacial relief resulting from Pleistocene glaciations, and has a temperate warm transition climate due to the clash of continental and oceanic climates. According to the Institute of Meteorology and Water Management State Research Institute (IMGW), in this area in 2018, the average annual air temperature was +9.0 °C, with August being the warmest month (+20.4 °C), and February the coldest (−4.1 °C). The sunshine duration ranged from 240 h in July to 40 h in January. The growing season spanned approximately 206 days, and snow cover persisted for 70 days. The annual sum of precipitation was 550 mm, with the highest amount of precipitation being recorded in July (140.7 mm) and the lowest in February (2.0 mm). The average air humidity was approximately 81%.

The rhizosphere of the three following plant species was chosen for the study: winter wheat (*Triticum aestivum*) of Julius cultivar, winter rape (*Brassica napus*) of Garou cultivar, and field pea (*Pisum sativum* subsp. *arvense*) of Milwa cultivar. The plants were grown simultaneously on three plots, each with a surface area of 20,000 m^2. The fraction size distribution of the soil used in the experiment was established and is presented in Table 1.

Table 1. The granulometric composition of the soil used in the experiment.

Plants	Particle Diameter, mm				Granulometric Subgroups
	< 0.002	0.002–0.020	0.020–0.050	0.050–2.0	
	%				
Triticum aestivum	3.9 [a]	18.1 [a]	17.7 [a]	60.2 [a]	sandy loam
Brassica napus	3.8 [a]	16.9 [b]	15.0 [b]	64.4 [a]	sandy loam
Pisum sativum ssp. *arvense*	4.0 [a]	16.0 [b]	14.0 [b]	66.0 [a]	sandy loam

Homogeneous groups denoted with letters (a, b) were calculated separately for each of the properties.

Winter rape was used as the previous crop of winter wheat, whereas winter wheat was used as the previous crop of winter rape and field pea. Winter wheat was sown on October 2, 2017, and the crop was harvested on August 2, 2018; winter rape was sown on August 19, 2017, and the crop was harvested on July 20, 2018, while pea was sown on April 7, 2018, and the crop was harvested on July 15, 2018. All plants were grown in accordance with the recommended technology for cultivating these plant species. All soil samples were collected the next day after the last crop harvest. In the case of winter wheat, fertilization was as follows: 66 kg N ha^{-1}, 20 kg P ha^{-1}, and 36 kg K ha^{-1}. The nitrogen dose was divided into three portions that were administered at stage 23 of plant development according to the BBCH scale, in the form of an aqueous solution of NH_4NO_3 and $CO(NH_2)_2 + H_2O$, and at stages 32 and 52 according to the BBCH scale, in the form of NH_4NO_3. Potassium and phosphorus were applied once before sowing. Potassium was used in the form of KCl and phosphorus in the form of $Ca(H_2PO_4)_2$. In the case of winter rape, fertilization was as follows: 75.8 kg N ha^{-1}, 40 kg P ha^{-1}, and 60 kg K ha^{-1}. Nitrogen dose was divided into three portions that were administered before sowing in the form of $CO(NH_2)_2$ and at stages 30 and 50 of plant development according to the BBCH scale, in the form of NH_4NO_3. Phosphorus and potassium were applied once before sowing in the form of $Ca(H_2PO_4)_2$ and KCl, respectively. No fertilization was applied in the case of field pea. The plant density of winter wheat per 1 m^2 was 400 plants; that of winter rape was 45 plants, and that of field pea, 90 plants. Soil was cultivated in a conventional (ploughing) manner in the case of all of the studied plant species.

The area of 20,000 m^2 was divided into five plots, each measuring 4000 m^2. After the harvest, 15 soil samples were collected at random from a depth of 0–20 cm from each plot using the zigzag sampling method. All soil samples collected from each plot were homogenized and combined into one collective sample. Hence, a total of five samples of soil from the cultivation of each plant species studied were used for microbiological and biochemical analyses. Soil samples were collected using an Enger-Riehm probe.

2.2. Methodology of Microbiological Analyses

2.2.1. Bacterial Count

Counts of organotrophic bacteria (Org) and actinobacteria (Act) were determined in individual soil samples from the cultivation of each of the studied plant species (winter wheat, winter rape, field pea) with the serial dilution method, in three replications. Microbial counts were performed according the media and procedure described by Borowik et al. [50]. The composition of the microbiological media was as follows: organotrophic bacteria (Bunt and Rovira medium): agar medium (peptone 1.0 g, yeast extract 1.0 g, $(NH_4)_2SO_4$ 0.5 g, $CaCl_2$, K_2HPO_4 0.4 g, $MgCl_2$ 0,2 g, $MgSO_4$ $7H_2O$ 0.5 g, salt Mo 0.03 g, $FeCl_2$ 0.01 g, agar 20.0 g, soil extract 250 cm^3, distilled water 750 cm^3, pH 6.6–7.0; Actinomycetes (Parkinson medium): soluble starch 10.0 g; casein 0.3 g; KNO_3 2.0 g; NaCl 2.0 g; K_2HPO_4 2.0 g; $MgSO4·7H2O$ 0.05 g; $CaCO_3$ 0.02 g; $FeSO_4$ 0.01 g; agar 20.0 g; H_2O 1 dm^3; 50 cm^3 aqueous solution of nystatin 0.05%; 50 cm^3 aqueous solution of actidione 0.05%; pH 7.0. Microorganisms were cultured on petri dishes at a temperature of 28 °C. The colony forming units (cfu) were counted every day for ten days using a colony counter. Counts of bacteria and actinobacteria isolated from the rhizosphere of particular plant species allowed us to compute the colony development index (CD), the ecophysiological diversity index (EP), and the microbial growth indexes at specific time intervals (Ks). Descriptions of the indexes and calculations are presented in De Leij et al. [51] and Tomkiel et al. 2015 [52]. The CD and EP were calculated from the following formulas [51]: CD = [N1/1 + N2/2 + N3/3 … N10/10] × 100, where N1, N2, N3, …N10 are the sum of ratios of the number of colonies of microorganisms identified on particular days (1, 2, 3, …10) to the total number of colonies identified throughout the study period, and EP = −Σ(pi·log10 pi), where pi is the ratio of the number of colonies of microorganisms identified on particular days to the total number of colonies identified throughout the study period. KS was determined with the use of the following formula [52]: Ks = (Nx/Nt) × 100, where Ks is the percentage of microbes cultured at specific time intervals, Nx is the number of colonies cultured at two-day intervals counted for ten days, and Nt is the total number of colonies cultured within ten days.

2.2.2. DNA Extraction and Bioinformatic Analysis of Specific Bacterial Taxa

DNA was isolated from the rhizosphere of the three plant species using a "Genomic Mini AX Bacteria+" kit. According to the manufacturer's instructions, the PCR reaction was carried out using Q5 Hotstart Hight-Fidelity DNA Polymerase (NEBNext). In the case of amplicon libraries, the data were pooled and normalized in the final stage of the library preparation. Afterwards, a 16S amplicon sequencing encoding gene was conducted for each DNA sample based on the hypervariable region V3–V4. The selected region was amplified and the library was developed using specific sequences of 341F (5′-CCTACGGGNGGCWGCAG-3′) and 785R (5′-GACTACHVGGGTATCTAATCC-3′) primers. The 16S library sequencing was performed on a MiSeq Reporter sequencer ver. 2.6 in the paired-end (PE) technology, 2 × 250 bp, using a v2 Illumina kit. The Illumina 16S Metagenomics workflow with MiSeq Reporter (MSR) software (San Diego, CA, USA) and Greengenes v13_5 software (South San Francisco, CA, USA) were used [53]. Preparation of the reference database included filtering low-quality, degenerate, and incomplete sequences, and then combining paired sequences based on the reference sequence database. The algorithm Uclust was assigned taxonomy, taking into account the ChimeraSlayer algorithm. Sequencing was performed by the Genomed S.A. Company (Warsaw, Poland).

2.3. Methodology of Biochemical Analyses

The activities of dehydrogenases (EC 1.1), catalase (EC 1.11.1.6), alkaline phosphatase (EC 3.1.3.1), acid phosphatase (EC 3.1.3.2), arylsulfatase (EC 3.1.6.1), β-glucosidase (EC 3.2.1.21), and urease (EC 3.5.1.5) were determined in triplicate in individual soil samples from the cultivation of each plant species studied. A detailed procedure of enzymatic activity determination is provided by Borowik et al. [50] and Borowik et al. [13]. The substrates used to determine the enzymatic activity included aqueous solutions of the following chemical compounds: 2,3,5-triphenyl tetrazolium chloride (TTC) for dehydrogenases, urea for urease, disodium 4-nitrophenyl phosphate hexahydrate (PNP) for phosphatases, potassium-4-nitrophenylsulfate (PNS) for arylsulfatase, and 4-nitrophenyl-β-D-glucopyranoside (PNG) for β-glucosidase. The activities of all enzymes except for catalase were determined by measuring the reaction product extinction using a Perkin-Elmer Lambda 25 spectrophotometer (Massachusetts, USA). Catalase activity was analyzed with the titration method. The activity of dehydrogenases was expressed in μmol TFF (tri-fenylformazane), that of catalase in mol O_2, that of alkaline phosphatase, acid phosphatase, arylsulfatase, and β-glucosidase in mmol PN (p-nitrophenol), and that of urease in mmol N-NH$_4$ kg^{-1} soil d.m. h^{-1}.

2.4. Methodology of Chemical and Physicochemical Analyses of Soil

Fraction size of soil from cultivation of winter wheat, winter rape, and field pea was measured using a Malvern Mastersizer 2000 Laser Diffraction. The physicochemical analyses included the determination of soil pH in 1 mol KCl dm^{-3} [54], hydrolytic acidity (HAC), and the sum of exchangeable base cations (EBC) according to the method outlined by Carter and Gregorich [55], whereas the chemical analyses included the determination of organic carbon content according to the method outlined by Tiurin [56], total nitrogen content according to the method outlined by Kjeldahl [57], available phosphorus and potassium contents according to the method outlined by Egner et al. [58], magnesium content with the atomic absorption spectrometry (AAS) according to the method outlined by Schlichting et al. [59], and exchangeable cations, K$^+$, Ca^{2+}, Mg^{2+}, and Na$^+$ according to the ISO 11260 [60] procedure. Determinations were carried out in the following solutions: HAC in 1 mol (CH$_3$COO)$_2$Ca dm^{-3}, EBC in 0.1 mol HCl dm^{-3}; organic carbon in a mixture of 0.13 mol K$_2$Cr$_2$O$_7$ and concentrated H$_2$SO$_4$ in the ratio of 1,1; total nitrogen—wet mineralization in concentrated H$_2$SO$_4$; available phosphorus and potassium in a mixture of 0.03 (CH$_3$CHOHCOO)$_2$Ca·H$_2$O and 0.02 mol HCl; magnesium in 0.012 mol CaCl$_2$·6H$_2$O; and exchangeable cations: K$^+$, Ca^{2+}, Mg^{2+}, and Na$^+$ in 1 mol CH$_3$COONH$_4$.

2.5. Statistical Analysis

Counts of microorganisms, the activity of soil enzymes, and the physicochemical and chemical properties of soil were developed statistically using the Statistica 13.1 package (StatSoft, Tulsa, OK, USA) [61]. The results were compared with ANOVA and then with a post hoc Tukey test (HSD). Homogenous groups were computed at $p = 0.05$. The data meet assumptions of normality and similar variance. Analyses of metagenomic profiles were performed on the STAMP 2.1.3. software (Australian Centre for Ecogenomics, School of Chemistry and Molecular Biosciences, The University of Queensland, St Lucia, QLD, Australia) [62] and the Circos 0.68 package (Canada's Michael Smith Genome Sciences Center, Vancouver, British Columbia V5Z 4S6, Canada) [63]. The abundance of each family and genus is directly proportional to the width of each band connecting bacterial taxa with an appropriate soil sample from the cultivation of: B—*Brassica napus*, T—*Triticum aestivum*, and P—*Pisum sativum* ssp. *arvense*. A specified color is assigned to each family and genus of bacteria. The outer ring represents the total percentage of 16S sequences, whereas the inner ring represents the number of 16S amplicon sequences assigned to a given taxon.

The relative abundance of bacteria was calculated using a two-sided test of statistical hypotheses, i.e., the G-test (w/Yates') + Fisher's, with the method of intervals confidence Asymptotic with CC [62]. The relative abundance of bacteria was visualized with the use of sequences whose percentage contribution was higher than 1%.

3. Results

3.1. Physicochemical and Chemical Properties of Soil

The pH value of soil from the cultivation of winter wheat, winter rape, and field pea ranged from 5.4 to 6.1 (Table 2). The most favorable pH value, the highest sorption capacity, and saturation with base cations were demonstrated for the soil sown with winter wheat, which offered the most beneficial conditions for the development of this crop (Table 3). This soil was also the richest in available phosphorus and potassium, and had the closest-to-optimal carbon to nitrogen ratio (14.8).

Table 2. The soil acidity and cation exchange capacity.

Plants	pH_{KCl}	HAC	EBC	CEC	BS %
		mmol (+) kg^{-1} d.m. of soil			
Triticum aestivum	6.1 [a]	19.4 [b]	174.8 [a]	194.2 [a]	90.0 [a]
Brassica napus	5.4 [b]	21.2 [a]	64.0 [b]	85.2 [b]	75.1 [b]
Pisum sativum ssp. *arvense*	5.6 [b]	19.2 [b]	46.7 [c]	65.9 [c]	70.9 [c]

Homogeneous groups denoted with letters (a, b, c) were calculated separately for each property.

Table 3. The content of carbon, phosphorus, nitrogen, potassium, sodium, calcium, and magnesium in soil.

Plants	Content		Available Forms			Interchangeable Forms			
	N_{total}	C_{total}	P	K	Mg	K	Ca	Na	Mg
	g kg^{-1} d.m. of Soil				mg kg^{-1} d.m. of Soil				
Triticum aestivum	1.0 [a]	14.8 [a]	69.1 [a]	224.1 [a]	53.0 [b]	272.0 [a]	500.0 [b]	20.0 [a]	83.3 [b]
Brassica napus	0.8 [c]	13.8 [b]	59.4 [b]	128.7 [b]	45.0 [c]	184.0 [b]	533.3 [a]	20.0 [a]	59.5 [c]
Pisum sativum ssp. *arvense*	0.9 [b]	14.3 [ab]	43.1 [c]	107.9 [c]	62.0 [a]	176.0 [b]	350.0 [c]	20.0 [a]	92.9 [a]

EBC—exchangeable base cations, HAC—hydrolytic activity, CEC—cation exchange capacity, BS—base saturation. Homogeneous groups denoted with letters (a, b, c) were calculated separately for each property.

3.2. Counts and Diversity of Microorganisms

The best conditions for microorganism proliferation were offered by the soil from field pea cultivation (Figure 1). The number of organotrophic bacteria in the field pea rhizosphere was 30% higher than in the rhizospheres of winter wheat and winter rape, whereas the number of actinobacteria was 98% and 110% higher than in the respective rhizospheres. The total bacteria count in the rhizosphere of winter wheat, winter rape, and field pea was not necessarily reflected in the proliferation rate and diversity of particular microorganisms. The analysis of the colony development index (CD) values demonstrated the fastest development of organotrophic bacteria in the rhizosphere of filed pea, and the fastest development of actinobacteria in the rhizosphere of winter wheat. Within the first four days, as many as 87% of the total organotrophs and 49% of the total actinobacteria grew in the soil from winter wheat cultivation (Figure 2). The values of the ecophysiological diversity index (EP) determined for organotrophic bacteria ranged from 0.887 (winter wheat rhizosphere) to 0.715 (field pea rhizosphere), and those calculated for actinobacteria from 0.843 (field pea rhizosphere) to 0.741 (winter rape rhizosphere). Analyses of CD and EP values allowed us to conclude that regardless of the plant species, higher values of both indices were determined for organotroph bacteria than for actinobacteria (CD = 34.27 vs. CD = 22.50; EP = 0.82 vs. EP = 0.79, respectively).

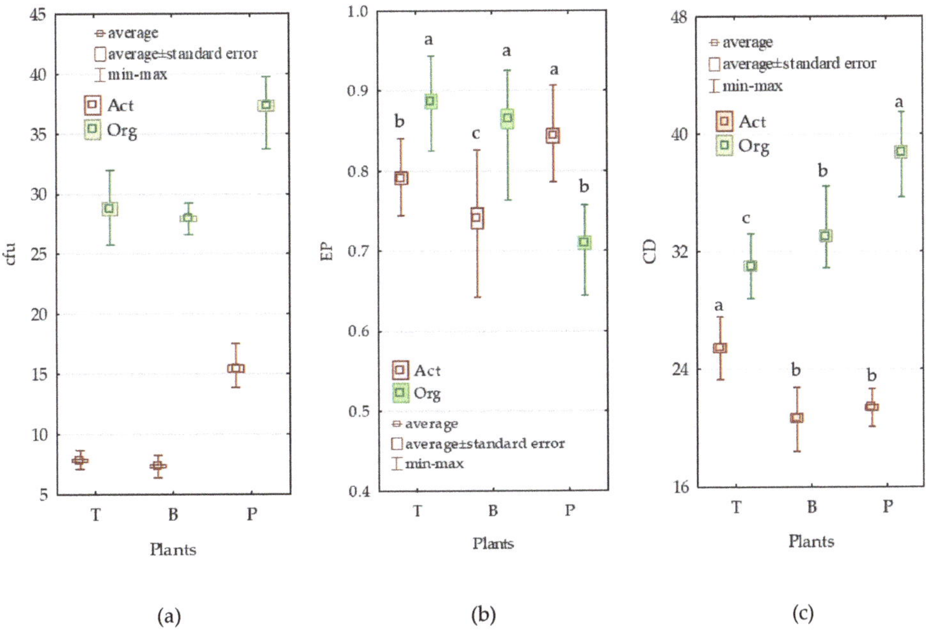

(a)   (b)   (c)

Figure 1. Microbiological properties of soil sown *Triticum aestivum* (T), *Brassica napus* (B), *Pisum sativum* ssp. *arvense* (P); (**a**) count of soil bacteria; (**b**) physiological diversity index of bacteria (EP); (**c**) colony development index (CD). Homogeneous groups denoted with letters (a, b, c) were calculated separately for each microorganism. Org—organotrophic bacteria, Act—actinobacteria.

The crop and its cultivation regimes have a significant impact on the soil microbiome. The prevailing phylum in the rhizosphere of all plants turned out to be *Proteobacteria*, which accounted for 33.05% in the soil from winter rape cultivation and 35.24% in the soil from winter wheat (Figure 3). Other phyla identified in all soils were *Actinobacteria* and *Firmicutes*. The greatest differences in OTU numbers were determined in the case of phylum *Actinobacteria*, i.e., the OTU number in the soil from field pea cultivation was higher by 6.39% than in the soil from winter wheat cultivation, whereas the OTU number in the soil from winter rape cultivation was higher by 4.87% than in the soil from winter wheat cultivation.

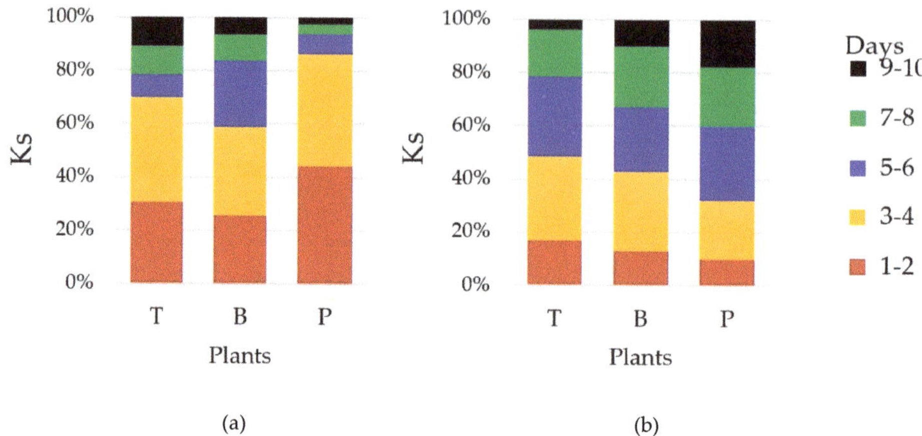

Figure 2. The growth and development index (Ks); (**a**) organotrophic bacteria; (**b**) actinobacteria.

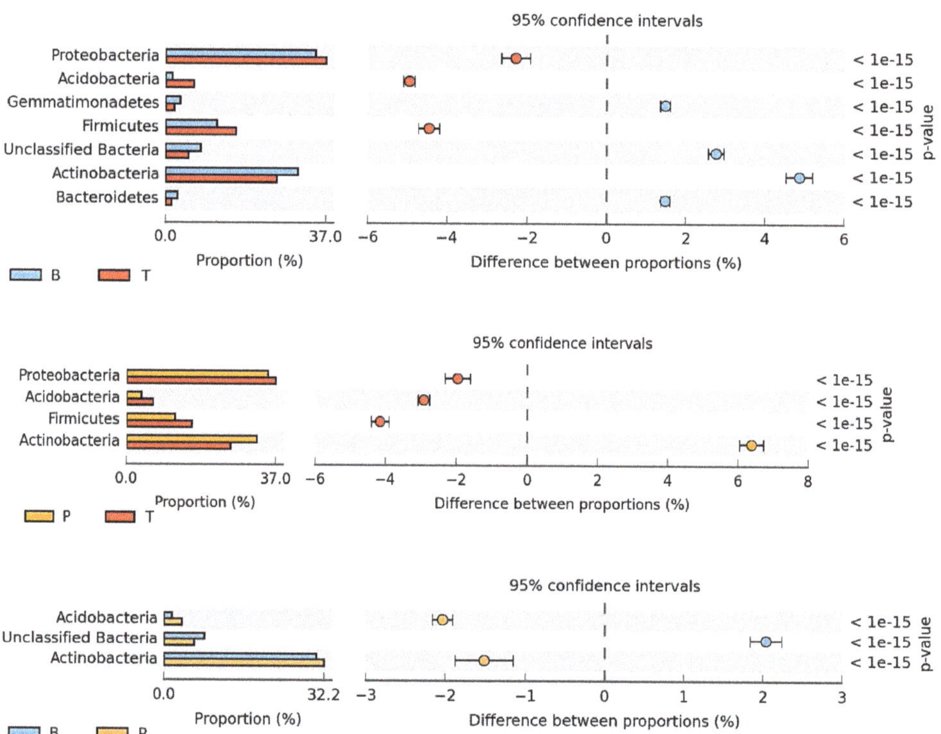

Figure 3. The relative abundance of dominant bacterial phyla in soil. Data on the number of readings greater than 1% of all OTUs. B—*Brassica napus*; T—*Triticum aestivum*; P—*Pisum sativum* ssp. *arvense*.

When comparing the effects of the discussed plant species on individual bacterial classes, it was found that the number of *Actinobacteria* OTUs determined in the field pea rhizosphere was higher by 6.82% than in the rhizosphere of winter wheat, and by 3.15% compared to the rhizosphere of winter rape (Figure 4). The second largest phylum was that of *Alphaproteobacteria*, whose OTU abundance was similar in soils from the cultivation of all of the studied plants.

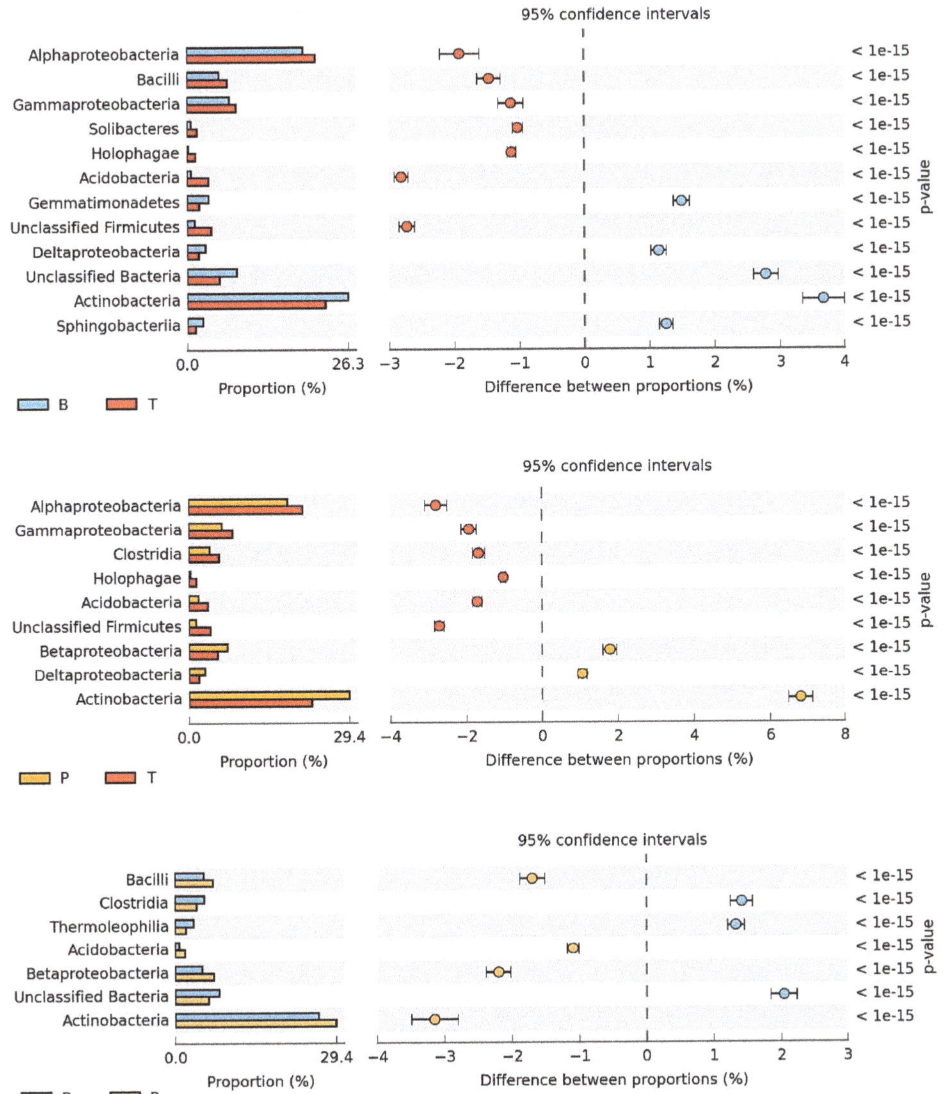

Figure 4. The relative abundance of dominant class bacteria in soil. Data on the number of readings greater than 1% of all OTUs. B—*Brassica napus*; T—*Triticum aestivum*; P—*Pisum sativum* ssp. *arvense*.

The greatest OTU abundance in order of rank, on average, and regardless of the cultivated plant species, was as follows: *Actinomycetales* (34,738 OTU) classified in the class *Actinobacteria*, phylum Actinobacteria; *Bacillales* (8083 OTU) classified in the class *Bacilli*, phylum *Firmicutes*; *Sphingomonadales* (11846 OTU); *Rhizobiales* (6897 OTU); *Rhodospirillales* (5897 OTU) classified in the class *Alphaproteobacteria*, phylum *Proteobacteria*; and for *Xanthomonadales* (5154) classified in the class *Gammaproteobacteria*, phylum *Proteobacteria* (Figure 5). Differences in the OTU abundance of particular classes in the soils from the cultivation of field pea, winter wheat, and winter rape usually did not exceed 3%. An exception was the order *Actinomycetales*, whose OTU number in the soil from field pea cultivation was higher by 6.94% than in the soil from winter wheat cultivation, and by 3.25% than in the soil from winter rape cultivation.

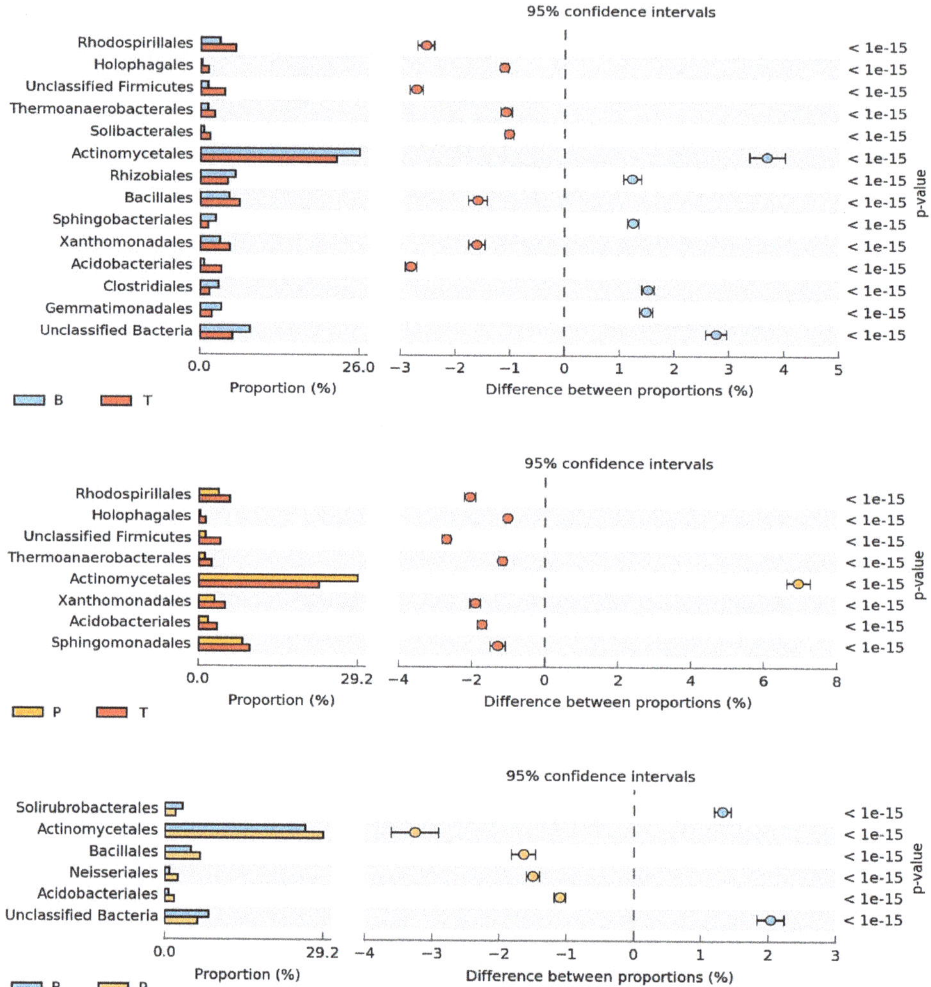

Figure 5. The relative abundance of dominant order bacteria in soil. Data on the number of readings greater than 1% of all OTUs. B—*Brassica napus*; T—*Triticum aestivum*; P—*Pisum sativum* ssp. *arvense*.

The cultivation of plants modified the soil microbiome also at the family level (Figure 6). In the soil sown with winter wheat, the highest numbers of OTUs were found for the following families:

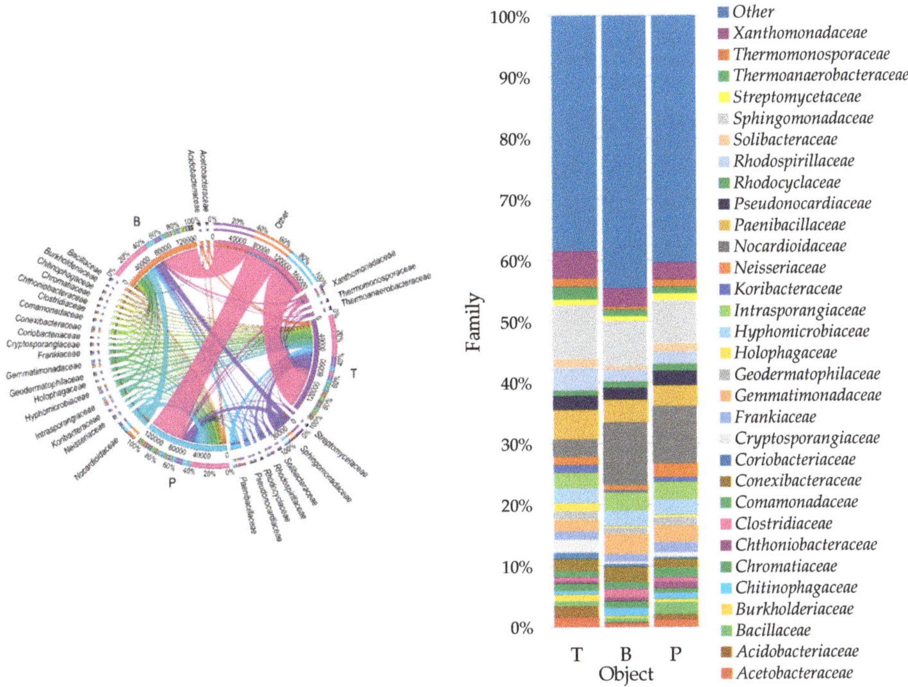

Figure 6. The relative abundance of dominant family bacteria in soil. Data on the number of readings greater than 1% of all OTUs. B—*Brassica napus*; T—*Triticum aestivum*; P—*Pisum sativum* ssp. *arvense*.

Paenibacillaceae (4.81%), *Xanthomonadaceae* (4.43%), and *Rhodospirillaceae* (3.71%), and in the soil from winter rape cultivation, for the families: *Nocardioidaceae* (10.37%), *Sphingomonadaceae* (7,40%), *Paenibacillaceae* (3,70%), *Gemmatimonadaceae* (3,31%), *Intrasporangiaceae* (3.06%), and *Xanthomonadaceae* (3.01%); and finally, in the soil from field pea cultivation, for the families: *Nocardioidaceae* (9.36%), *Sphingomonadaceae* (7.08%), and *Paenibacillaceae* (3.43%). At this taxonomic level, the greatest differences in the effects of individual plant species were noticeable in the abundance of OTUs from the *Nocardioidaceae*.

In the soils from the cultivation of winter rape and field pea, there were by 7.93% and 6.91% more OTUs, respectively, than in the soil from winter wheat cultivation. It is worthy of notice that the highest number of genera were classified in the soil sown with winter wheat, and the lowest in the soil sown with winter rape (Figure 7). In the soils from winter rape and field cultivation, the prevailing genus turned out to be the *Nocardioides*, which accounted for 7.45% and 6.63% of all identified bacteria, respectively, whereas in the soil from winter wheat cultivation, it was the *Kaistobacter* genus (6.32%).

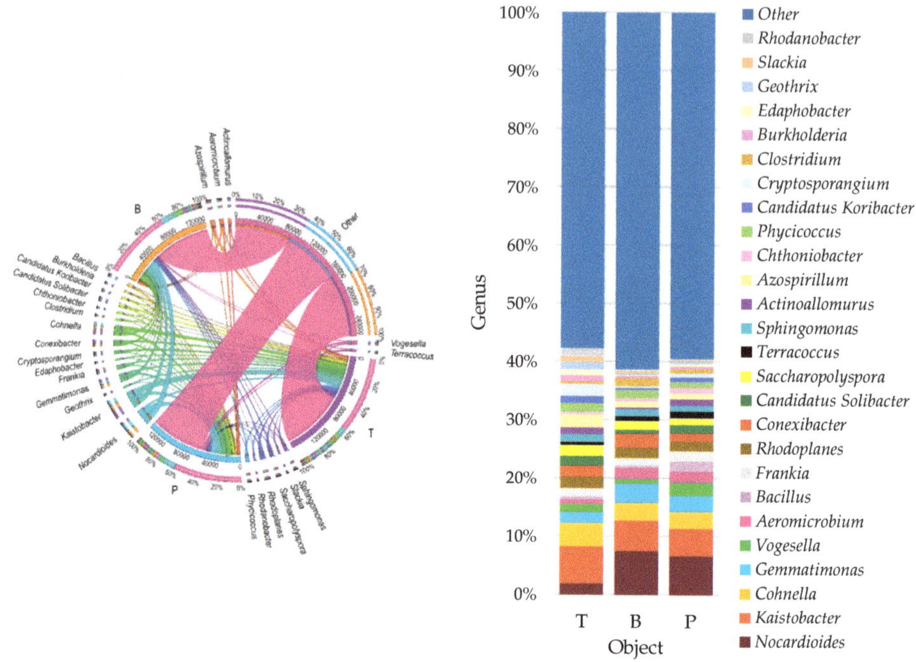

Figure 7. The relative abundance of dominant genus bacteria in soil. Data on the number of readings greater than 1% of all OTUs. B—*Brassica napus*; T—*Triticum aestivum*; P—*Pisum sativum* ssp. *arvense*.

3.3. Enzymatic Activity of Soil

Apart from the numbers and diversity of microorganisms, the biological activity of soil and, thus, its fertility, is determined by the activity of soil enzymes (Table 4). In the present study, the enzymatic activity of soil was significantly affected by the plant crop and its cultivation regimes. The highest activities of dehydrogenases and acid phosphatase were determined in the soil from field pea cultivation, and those of catalase, urease, alkaline phosphatase, β-glucosidase, and arylsulfatase in the soil from winter wheat cultivation. Especially big differences between plant species were noticeable in the activities of alkaline phosphatase and arylsulfatase, which, in the soil sown with winter wheat, were respectively 6.2 times and 2.7 times higher than in the soil from winter rape cultivation, and 3.5 times and 1.9 times higher than in the soil from field pea cultivation. This activity was strongly correlated with the productivity of individual plant species (Figure 8). Considering the total activity of all analyzed enzymes taking part in carbon, nitrogen, phosphorus, and sulfur metabolism, the highest biochemical activity was demonstrated for the rhizosphere of winter wheat, followed by that of field pea and winter rape.

The activities of catalase, urease, alkaline phosphatase, β-glucosidase, and arylsulfatase were positively significantly correlated with the sum of exchangeable base cations and with the degree of soil saturation with base cations (Figure 9). The sorption properties of soil had no effect on the activities of dehydrogenases and acid phosphatase. A similar response of the enzymes was observed regarding the contents of available and exchangeable phosphorus in the soil. The activities of dehydrogenases, urease, phosphatases (both, acid and alkaline), and arylsulfatase were also positively significantly correlated with total nitrogen content of the soil.

Table 4. The soil acidity and the cation exchange capacity.

Plants	Deh µM TFF M O$_2$	Cat mM PNP	Pal	Pac	Aryl	Glu	Ure mM N-NH$_4$
				mM PNP			
Triticum aestivum	5.67 [b]	3.45 [a]	2.41 [a]	3.29 [b]	0.51 [a]	1.16 [a]	2.07 [a]
Brassica napus	5.18 [c]	3.25 [b]	0.39 [C]	2.73 [c]	0.19 [c]	0.97 [b]	0.94 [C]
Pisum sativum ssp. *arvense*	6.16 [a]	3.49 [b]	0.69 [b]	3.49 [a]	0.27 [b]	0.84 [C]	1.47 [b]

Homogeneous groups denoted with letters (a, b, c) were calculated separately for each enzyme.

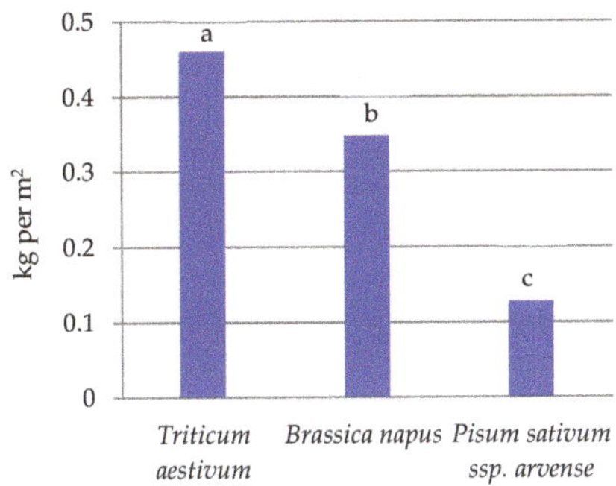

Figure 8. The yield of plants in kg per m^2.

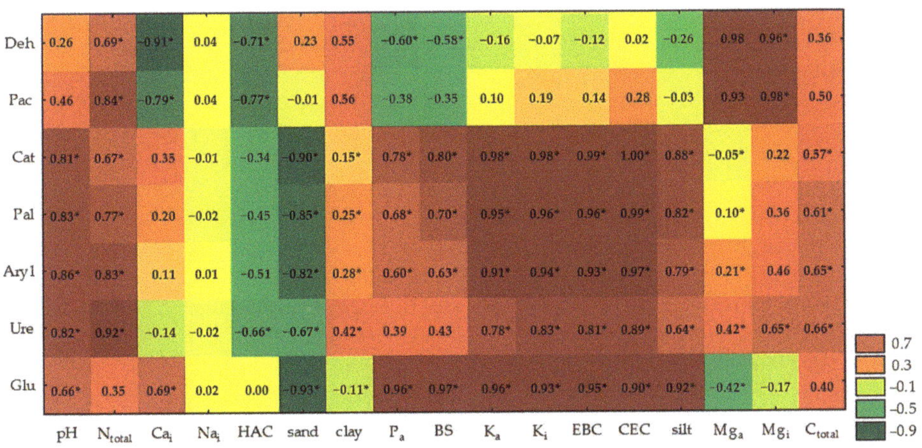

	pH	N$_{total}$	Ca$_i$	Na$_i$	HAC	sand	clay	P$_a$	BS	K$_a$	K$_i$	EBC	CEC	silt	Mg$_a$	Mg$_i$	C$_{total}$
Deh	0.26	0.69*	−0.91*	0.04	−0.71*	0.23	0.55	−0.60*	−0.58*	−0.16	−0.07	−0.12	0.02	−0.26	0.98	0.96*	0.36
Pac	0.46	0.84*	−0.79*	0.04	−0.77*	−0.01	0.56	−0.38	−0.35	0.10	0.19	0.14	0.28	−0.03	0.93	0.98*	0.50
Cat	0.81*	0.67*	0.35	−0.01	−0.34	−0.90*	0.15*	0.78*	0.80*	0.98*	0.98*	0.99*	1.00*	0.88*	−0.05*	0.22	0.57*
Pal	0.83*	0.77*	0.20	−0.02	−0.45	−0.85*	0.25*	0.68*	0.70*	0.95*	0.96*	0.96*	0.99*	0.82*	0.10*	0.36	0.61*
Aryl	0.86*	0.83*	0.11	0.01	−0.51	−0.82*	0.28*	0.60*	0.63*	0.91*	0.94*	0.93*	0.97*	0.79*	0.21*	0.46	0.65*
Ure	0.82*	0.92*	−0.14	−0.02	−0.66*	−0.67*	0.42*	0.39	0.43	0.78*	0.83*	0.81*	0.89*	0.64*	0.42*	0.65*	0.66*
Glu	0.66*	0.35	0.69*	0.02	0.00	−0.93*	−0.11*	0.96*	0.97*	0.96*	0.93*	0.95*	0.90*	0.92*	−0.42*	−0.17	0.40

Legend: 0.7, 0.3, −0.1, −0.5, −0.9

Figure 9. Pearson correlation coefficients between enzyme activity and soil physicochemical and chemical properties, $n = 15$, $p = 0.05$, * significant differences.

4. Discussion

4.1. Physicochemical and Chemical Properties of Soil

The abundance of available phosphorus and potassium in the soil, as well as the highest sorption capacity and soil saturation with base cations, were probably the main factors which affected the yields of the tested plants (Figure 8). Grain yield produced from 1 m^2 of the plot sown with winter wheat reached 0.462 kg grains, that achieved from the plot sown with winter rape reached 0.350 kg grains, and that for field pea reached only 0.130 kg grains. Next to the natural plant features, the physicochemical properties of the soil are closely related to the development of the root system. Besides the varied chemical composition of the plant species grown in the study, the size of the root system was probably the main determinant of the differences observed in the microbiological and biochemical properties of soil, because plant species and soil fertility are the key indicators of the biological activities of soil [64]. The superiority of one factor over another is affected by multiple habitat factors, like, e.g., soil tillage system [65] the contents of phosphorus and Ca^{2+}, Mg^{2+}, Al^{3+} cations, and the pH value of soil [66].

4.2. Counts and Diversity of Bacteria

The physicochemical properties of the soil from winter wheat cultivation should, theoretically, be more favorable to the development of organotrophic bacteria and actinobacteria than the soil used for winter rape and field pea cultivation. This is due to, among other things, more favorable fraction size distribution, a higher sorption capacity, and lower acidification of the soil. However, the experimental data also pointed to some other dependencies influenced by the species of the cultivated plant. In spite of the fact that the physicochemical properties of the soil sown with field pea were far from being the most favorable, it was its rhizosphere that was the most abundant in organotrophs and actinobacteria. This can be explained by the difference in the root system morphology between winter wheat and field pea, in favor of the latter, as well as by the capability of field pea roots for symbiosis with atmospheric nitrogen-binding bacteria from the genus *Rhizobium*, and by the difference in the chemical composition of root secretions. The aforementioned factors contributed to a more dynamic development of organotrophic bacteria of r strategy than of k strategy, which was indicated by the values of their colony development index (CD). Usually, the greater inflow of nutrients to the natural environment aids the development of fast-growing microorganisms [34]. The k strategists, i.e., the slow-growing bacteria, are more stable in this respect. They are responsible for maintaining soil homeostasis, and are also resistant to the adverse effects of environmental conditions [67], as well as being typical of soil ecosystems [15,48].

Generally, organotrophic bacteria proliferate faster in the soil than actinobacteria as a result of organic matter inflow to the natural environment [13]. Such a response of microorganisms is not always associated with their ecological diversity [48]. In the present study, the highest value of the ecophysiological diversity index (EP) of organotrophic bacteria was determined in the winter wheat rhizosphere (0.877) and that of actinobacteria in the field pea rhizosphere. In the case of the latter microorganisms, the EP values reached 0.843 in the soil from pea field cultivation, 0.791 in the soil from winter wheat cultivation, and 0.684 in the soil from winter rape cultivation.

Under natural conditions, microbiome stability is largely affected by the species of growing plant and, especially, by its root system [41,48]. By providing water-soluble compounds to plants, including organic acids, sugars, or amino acids, the roots of plants stimulate the microbiological activity of soil [68,69], whereas plant growth promoting bacteria (PGPR) colonize roots and increase the root system biomass [40,70]. The PGPR modify the root architecture through the production of phytohormones, siderophores, and hydrogen cyanide, as well as by nitrogen uptake and mechanisms of phosphates stabilization [15,47,71]. Therefore, the rhizosphere is characterized by greater diversity of the population of microorganisms than the soil distant from the root system of plants [42,46,72].

Another important factor which determines soil health, and thus, soil quality and productivity, is the structure of microbial communities [73], because these communities affect the stability of the soil ecosystem [7,11,74]. There is a strong correlation between the soil microbiome and the plant microbiome [75,76]. It is the soil bacteriome that often determines the quality features of cultivable plants [77]. According to Bakker et al. [78] and Xu et al. [76], the prevailing phyla in the arable soil include *Proteobacteria, Actinobacteria, Bacteroidetes, Firmicutes*, and *Acidobacteria*. According to Xu et al. [76], proteobacteria comprised 40% and 43% in the pot and the field experiments, respectively. Proteobacteria can quickly respond to nutrient changes in the rhizosphere. Maron et al. [67] and Pascault et al. [79] have emphasized that *Acidobacteria, Actinobacteria, Planctomycetes, Chloroflexi*, and *Gemmatimonadetes* are classified among slow-growing microorganisms (k-strategists), whereas *Proteobacteria* (mainly those from classes *Alphaproteobacteria* and *Gammaproteobacteria*) and *Firmicutes* are among the fast-growing ones (r-strategists).

The dependency between a plant species and the microbiological communities in the rhizosphere that was observed in the present study was also reported by Huang et al. [80]. According to Maron et al. [67], diminished diversity of microorganisms retards the degradation of both autochthonous and allochthonous carbon sources, thereby reducing the global emission of CO_2 by as much as 40%, whereas the importance of the diversity effect increases along with increasing the availability of nutrients. Bacteria from various families differ significantly in their genetics, which determine their functions in the soil environment [81]. For instance, bacteria representing the families *Solibacteraceae* and *Acidobacteriaceae* play an active part in protein and carbohydrate mineralization; those from the family *Baciliaceae* degrade chitin and cellulose, and participate in the biosynthesis of plant growth hormones and secondary metabolites [82]; bacteria from the family *Burkholderiaceae* are active participants of bioplastics biodegradation [83]; and *Streptomycetaceae*, *Pseudonocardiaceae*, and *Promicromonosporaceae* exhibit robust activity against carboxymethyl cellulose, xylan, chitin, and pectin substrates [84].

According to Chaparro et al. [66], the secretions of the roots of various species or even ecotypes of plants, which determine the soil microbiome, differ in their chemical compositions. They manifest plant responses to the chemical signals emitted by soil microorganisms. Their secretion process can proceed both at the passive and active transport pathways. Root secretions determine interactions between plants and the soil microbiome, and therefore, regardless of the soil type, the crop and its cultivation regime is the key determinant of the soil microbiome [64]. The above factors contribute to a more favorable structure in the bacteriome of the soil from winter wheat cultivation compared to the soils from winter rape and field pea cultivation.

4.3. Enzymatic Activity of Soil

The results of ample investigations [10,85–87] have demonstrated that the analysis of the activity of soil enzymes is of great importance to the evaluation of soil quality and productivity, due to its high sensitivity, ease of measurements, and a high correlation with plant yield. The enzymatic activity of soil is usually strongly associated with soil colonization by microorganisms. In the reported study, this effect was observed only in the case of dehydrogenases and acid phosphatase. Their activities were highest in the soil from cultivation of field pea, i.e., in the rhizosphere had the greatest effect on the promotion of soil colonization by cultivable bacteria (organotrophic bacteria and actinobacteria). The positive correlation between the abundance of microbial communities and the activity of dehydrogenases has been shown by many authors [88,89], who explained this phenomenon by the localization of these enzymes in viable microorganism cells [86]. According to Merino et al. [90], intracellular, rather than extracellular, enzymes provide information about the potential activity of a community of soil microorganisms. In the present study, the activities of the other tested enzymes, in particular of urease, alkaline phosphatase, β-glucosidase, and arylsulfatase, were higher in the soil sown with winter wheat than in those of field pea and winter rape. The activities of extracellular enzymes are positively correlated with the sorption capacity of soil [91], whereas the soil from the cultivation of winter wheat was characterized by the best physicochemical properties. The activities of these enzymes may, however, be suppressed by the excessive sorption capacity caused by, e.g., biocarbon supplementation [92].

5. Conclusions

The crop and its cultivation regimes contributed to the development of specific conditions which modify soil microbiome. The rhizosphere of winter wheat had a more beneficial effect on bacteria development and enzyme activity than those of winter rape and field pea, as indicated by the values of the ecophysiological diversity index (EP) in the soils sown with the tested plants. Microbiological indices, including bacterial count and diversity, as well activities of soil enzymes, are reliable indicators of soil environment conditions, and are also helpful parameters in the evaluation of soil fertility and productivity.

Author Contributions: J.W. conceived and designed the ideas and wrote the manuscript with the help of A.B., J.K., J.O.; J.O. conducted the field experiment; J.W.; J.K.; A.B. collected and analyzed data, and wrote the manuscript; A.B. conducted the bioinformatic analysis and visualization of data; all authors contributed to the final version of this manuscript.

Funding: This study was supported by the Ministry of Science and Higher Education funds for statutory activity. Project financially supported by Minister of Science of Higher Education in the range of the program entitled "Regional Initiative of Excellence" for the years 2019-2022, Project No. 010/RID/2018/19, amount of funding 12.000.000 PLN."

Conflicts of Interest: The authors declare no conflict of interest. The funders had no role in the design of the study; in the collection, analyses, or interpretation of data; in the writing of the manuscript, or in the decision to publish the results.

References

1. Congreves, K.A.; Hayes, A.; Verhallen, E.A.; Van Eerd, L.L. Long-term impact of tillage and crop rotation on soil health at four temperate agroecosystems. *Soil Tillage Res.* **2015**, *152*, 17–28. [CrossRef]
2. Lehman, M.R.; Cambardella, A.C.; Stott, E.D.; Acosta-Martinez, V.; Manter, K.D.; Buyer, S.J.; Maul, E.J.; Smith, L.J.; Collins, P.H.; Halvorson, J.J.; et al. Understanding and Enhancing Soil Biological Health: The Solution for Reversing Soil Degradation. *Sustainability* **2015**, *7*, 988. [CrossRef]
3. Baveye, P.C.; Baveye, J.; Gowdy, J. Soil "ecosystem" services and natural capital: Critical appraisal of research on uncertain ground. *Front. Environ. Sci.* **2016**, *4*, 1–49. [CrossRef]
4. Doran, J.W. Soil health and global sustainability: Translating science into practice. *Agric. Ecosyst. Environ.* **2002**, *88*, 119–127. [CrossRef]
5. Cardoso, E.J.B.N.; Vasconcellos, R.L.F.; Bini, D.; Miyauchi, M.Y.H.; dos Santos, C.A.; Alves, P.R.L.; de Paula, A.M.; Nakatani, A.S.; Pereira, J.M.; Nogueira, M.A. Soil health: Looking for suitable indicators. What should be considered to assess the effects of use and management on soil health? *Sci. Agric.* **2013**, *70*, 274–289. [CrossRef]
6. Legaz, B.V.; Maia De Souza, D.; Teixeira, R.F.M.; Antón, A.; Putman, B.; Sala, S. Soil quality, properties, and functions in life cycle assessment: An evaluation models. *J. Clean. Prod.* **2017**, *140*, 502–515. [CrossRef]
7. Bünemann, E.K.; Bongiorno, G.; Bai, Z.; Creamer, R.E.; De Deyn, G.; de Goede, R.; Fleskens, L.; Geissen, V.; Kuyper, T.W.; Mäder, P.; et al. Soil quality—A critical review. *Soil Biol. Biochem.* **2018**, *120*, 105–125. [CrossRef]
8. Veum, K.S.; Sudduth, K.A.; Kremer, R.J.; Kitchen, N.R. Sensor data fusion for soil health assessment. *Geoderma* **2017**, *305*, 53–61. [CrossRef]
9. Nannipieri, P.; Ascher, J.; Ceccherini, M.T.; Landi, L.; Pietramellara, G.; Renella, G. Microbial diversity and soil functions. *Eur. J. Soil Sci.* **2003**, *54*, 655–670. [CrossRef]
10. Wyszkowska, J.; Borowik, A.; Kucharski, M.; Kucharski, J. Applicability of biochemical indices to quality assessment of soil pulluted with heavy metal. *J. Elem.* **2013**, *18*, 723–732. [CrossRef]
11. Compant, S.; Duffy, B.; Nowak, J.; Clement, C.; Barka, A. Use of plant growth-promoting bacteria for biocontrol of plant diseases: Principles, mechanisms of action, and future prospects. *Appl. Environ. Microbiol.* **2005**, *71*, 4951–4959. [CrossRef] [PubMed]
12. Zaborowska, M.; Woźny, G.; Wyszkowska, J.; Kucharski, J. Biostimulation of the activity of microorganisms and soil enzymes through fertilisation with composts. *Soil Res.* **2018**, *56*, 737–751. [CrossRef]
13. Borowik, A.; Wyszkowska, J. Remediation of soil contaminated with diesel oil. *J. Elem.* **2018**, *23*, 767–788. [CrossRef]
14. Zaborowska, M.; Kucharski, J.; Wyszkowska, J. Biochemical and microbiological activity of soil contaminated with o-cresol and biostimulated with *Perna canaliculus* mussel meal. *Environ. Monit. Assess.* **2018**, *190*, 602. [CrossRef]
15. Sharifi, R.; Ryu, C.M. Sniffing bacterial volatile compounds for healthier plants. *Curr. Opin. Plant Biol.* **2018**, *44*, 88–97. [CrossRef]

16. Bell, C.W.; Acosta-Martinez, V.; McIntyre, N.E.; Cox, S.; Tissue, D.T.; Zak, J.C. Linking microbial community structure and function to seasonal differences in soil moisture and temperature in a Chihuahuan Desert Grassland. *Microb. Ecol.* **2009**, *58*, 827–842. [CrossRef]

17. Borowik, A.; Wyszkowska, J. Soil moisture as a factor affecting the microbiological and biochemical activity of soil. *PlantSoil Environ.* **2016**, *62*, 250–255. [CrossRef]

18. Gordon, H.; Haygarth, P.M.; Bardgett, R.D. Drying and rewetting effects on soil microbial community composition and nutrient leaching. *Soil Biol. Biochem.* **2008**, *40*, 302–311. [CrossRef]

19. Singurindy, O.; Molodovskaya, M.; Richards, B.K.; Steenhuis, T.S. Nitrous oxide emission at low temperatures from manure-amended soiils under corn (*Zea mays* L.). *Agric. Ecosyst. Environ.* **2009**, *132*, 74–81. [CrossRef]

20. Craine, J.; Spurr, R.; McLauchlan, K.; Fierer, N. Lanscape-level variation in temperature sensitivity of soil organic carbon decomposition. *Soil Biol. Biochem.* **2010**, *42*, 373–375. [CrossRef]

21. Silva, C.C.; Guido, M.L.; Ceballos, J.M.; Marsch, R.; Dendooven, L. Production of carbon dioxide and nitrous oxide in alkaline saline soil of Texcoco at different water contents amended with urea: A laboratory study. *Soil Biol. Biochem.* **2008**, *40*, 1813–1822. [CrossRef]

22. Asuming-Brempong, S.; Ganter, S.; Adiku, S.G.K.; Archer, G.; Edusei, V.; Tiedje, J.M. Changes in the biodiversity of microbial populations in tropical soils under different fallow treatments. *Siol Biol. Biochem.* **2008**, *40*, 2811–2818. [CrossRef]

23. Garcia-Ruiz, R.; Ochoa, V.; Hinojosa, M.B.; Carreira, J.A. Suitability of enzyme activities for the monitoring of soil quality improvement in organic agricultural systems. *Soil Biol. Biochem.* **2008**, *40*, 2137–2145. [CrossRef]

24. Udawatta, R.P.; Kremer, R.J.; Garrett, H.E.; Anderson, S.H. Soil enzyme activities and physical propertie4s in a watershed managed under agroforestry and row-crop systems. *Agric. Ecosyst. Environ.* **2009**, *131*, 98–104. [CrossRef]

25. Mandal, A.; Patra, A.K.; Singh, D.; Swarup, A.; Masto, R.E. Effect of long-term application of manure and fertilizer on biological and biochemical activities in soil during corp development stages. *Bioresour. Technol.* **2007**, *98*, 3585–3592. [CrossRef] [PubMed]

26. Feng, X.; Simpson, M.J. Temperature and substrate controls on microbial phospholipid fatty acid composition during incubation of grassland soils contrasting in organic matter quality. *Soil Biol. Biochem.* **2009**, *41*, 804–812. [CrossRef]

27. Liu, Z.; Fu, B.; Zheng, X.; Liu, G. Plant biomass, soil water content and soil N: P ratio regulating soil microbial functional diversity in a temperature steppe: A regional scale study. *Soil Biol. Biochem.* **2010**, *42*, 445–450. [CrossRef]

28. Boros-Lajszner, E.; Wyszkowska, J.; Kucharski, J. Use of zeolite to neutralise nickel in a soil environment. *Environ. Monit. Assess.* **2018**, *190*, 54. [CrossRef]

29. Wyszkowska, J.; Boros-Lajszner, E.; Borowik, A.; Kucharski, J.; Baćmaga, M.; Tomkiel, M. Changes in the microbiological and biochemical properties of soil contaminated with zinc. *J. Elem.* **2017**, *22*, 437–451. [CrossRef]

30. Kucharski, J.; Jastrzębska, E. Effects of heating oil on the count of microorganisms and physico-chemical properties of soil. *Pol. J. Environ. Stud.* **2005**, *14*, 195–204.

31. Lipińska, A.; Wyszkowska, J.; Kucharski, J. Diversity of organotrophic bacteria, activity of dehydrogenases and urease as well as seed germination and root growth *Lepidium sativum*, *Sorghum saccharatum* and *Sinapis alba* under the influence of polycyclic aromatic hydrocarbons. *Environ. Sci. Pollut. Res.* **2015**, *22*, 18519–18530. [CrossRef] [PubMed]

32. Wyszkowska, J.; Borowik, A.; Kucharski, J. The resistance of *Lolium perenne* L. × hybridum, *Poa pratensis*, *Festuca rubra*, F. *arundinacea*, *Phleum pratense* and *Dactylis glomerata* to soil pollution by diesel oil and petroleum. *PlantSoil Environ.* **2019**, *65*, 307–312. [CrossRef]

33. Baćmaga, M.; Wyszkowska, J.; Kucharski, J. Biostimulation as a process aiding tebuconazole degradation in soil. *J. Soil Sediment.* **2019**, *19*, 3728–3741. [CrossRef]

34. Kucharski, J.; Tomkiel, M.; Baćmaga, M.; Borowik, A.; Wyszkowska, J. Enzyme activity and microorganisms diversity in soil contaminated with the Boreal 58 WG herbicide. *J. Environ. Sci. Health B.* **2016**, *51*, 446–454. [CrossRef] [PubMed]

35. Tang, L.; Dong, J.; Ren, L.; Zhu, Q.; Huang, W.; Liu, Y.; Lu, D. Biodegradation of chlorothalonil by *Enterobacter cloacae* TUAH-1. *Int. Biodeter. Biodegr.* **2017**, *121*, 122–130. [CrossRef]

36. Hund-Rinke, K.; Simon, M. Bioavailability assessment of contaminants in soils via respiration and nitrification tests. *Environ. Pollut.* **2008**, *153*, 468–475. [CrossRef]

37. Tian, Y.; Zhang, X.; Liu, J.; Chen, Q.; Gao, L. Microbial properties of rhizosphere soils as affected by rotation, grafting, and soil sterilization in intensive vegetable production systems. *Sci. Hortic.* **2009**, *123*, 139–147. [CrossRef]

38. Kohler, J.; Caravaca, F.; Roldán, A. Effect of drought on the stability of rhizosphere soil aggregates of Lactuca sativa grown in a degreded soil inoculated with PGPR and AM fungi. *Appl. Soil Ecol.* **2009**, *42*, 160–165. [CrossRef]

39. Hai, L.; Li, X.G.; Suo, D.R.; Guggenberger, G. Long-term fertilization and manuring effects on physically-separated soil organic matter pools under a wheat-wheat-maize cropping system in an arid region of China. *Soil Biol. Biochem.* **2010**, *42*, 253–259. [CrossRef]

40. Carminati, A.; Schneider, C.L.; Moradi, A.B.; Zarebanadkouki, M.; Vetterlein, D.; Vogel, H.J.; Hildebrandt, A.; Weller, U.; Schüler, L.; Oswald, S.E. How the rhizosphere may favor water availability to roots. *Vadose Zone J.* **2011**, *10*, 988–998. [CrossRef]

41. Shaikh, S.; Wani, S.; Sayyed, R. Impact of Interactions between Rhizosphere and Rhizobacteria: A Review. *J. Bacteriol. Mycol.* **2018**, *5*, 1058.

42. Benard, P.; Zarebanadkouki, M.; Brax, M.; Kaltenbach, R.; Jerjen, I.; Marone, F.; Couradeau, E.; Felde, V.J.; Kaestner, A.; Carminati, A. Microhydrological niches in soils: How mucilage and EPS alter the biophysical properties of the rhizosphere and other biological hotspots. *Vadose Zone J.* **2019**, *18*, 1–10. [CrossRef]

43. Ahemad, M.; Kibret, M. Mechanisms and applications of plant growth promoting rhizobacteria: Current perspective. *J. King Saud Univ. Sci.* **2014**, *26*, 1–20. [CrossRef]

44. Kang, B.G.; Kim, W.T.; Yun, H.S.; Chang, S.C. Use of plant growth-promoting rhizobacteria to control stress responses of plant roots. *Plant Biotechnol. Rep.* **2010**, *4*, 179–183. [CrossRef]

45. Nannipieri, P.; Aschner, J.; Ceccherini, M.T.; Landi, L.; Pietramellara, G.; Renella, G.; Valori, F. Microbial diversity and microbial activity in the rhizosphere. *Ci Suelo (Argentina)* **2007**, *25*, 89–97.

46. Berendsen, R.L.; Pieterse, C.M.; Bakker, P.A. The rhizosphere microbiome and plant health. *Trends Plant Sci.* **2012**, *17*, 478–486. [CrossRef]

47. Hassan, M.K.; McInroy, J.A.; Kloepper, J.W. The Interactions of Rhizodeposits with Plant Growth-Promoting Rhizobacteria in the Rhizosphere: A Review. *Agriculture* **2019**, *9*, 142. [CrossRef]

48. Mueller, C.W.; Carminati, A.; Kaiser, C.; Subke, J.A.; Gutjahr, C. Rhizosphere functioning and structural development as complex interplay between plants, microorganisms and soil minerals. *Front. Environ. Sci.* **2019**, *7*, 130. [CrossRef]

49. Rausch, P.; Rühlemann, M.; Hermes, B.M.; Doms, S.; Dagan, T.; Dierking, K.; Domin, H.; Fraune, S.; von Frieling, J.; Hentschel, U.; et al. Comparative analysis of amplicon and metagenomic sequencing methods reveals key features in the evolution of animal metaorganisms. *Microbiome* **2019**, *7*, 133. [CrossRef]

50. Borowik, A.; Wyszkowska, J.; Wyszkowski, M. Resistance of aerobic microorganisms and soil enzyme response to soil contamination with Ekodiesel Ultra fuel. *Environ. Sci. Pollut. Res.* **2017**, *24*, 24346–24363. [CrossRef]

51. De Leij, F.A.A.; Whipps, J.M.; Lynch, J.M. The use of colony development for the characterization of bacterial communities in soil and on roots. *Microb. Ecol.* **1994**, *27*, 81–97. [CrossRef] [PubMed]

52. Tomkiel, M.; Baćmaga, M.; Wyszkowska, J.; Kucharski, J.; Borowik, A. The effect of carfentrazone-ethyl on soil microorganisms and soil enzymes activity. *Arch. Environ. Prot.* **2015**, *41*, 3–10. [CrossRef]

53. De Santis, T.Z.; Hugenholtz, P.; Larsen, N.; Rojas, M.; Brodie, E.L.; Keller, K.; Huber, T.; Dalevi, D.; Hu, P.; Andersen, G.L. Greengenes, a chimera-checked 16S rRNA gene database and workbench compatible with ARB. *Appl. Environ. Microbiol.* **2006**, *72*, 5069–5072. [CrossRef] [PubMed]

54. ISO 10390. *Soil Quality—Determination of pH*; International Organization for Standardization: Geneva, Switzerland, 2005.

55. Carter, M.R.; Gregorich, E.G. *Soil Sampling and Methods of Analysis*, 2nd ed.; CRC Press: Boca Raton, FL, USA, 2008; p. 1224.

56. Nelson, D.W.; Sommers, L.E. Total Carbon, Organic Carbon, and Organic matter. In *Method of Soil Analysis: Chemical Methods*; Sparks, D.L., Ed.; American Society of Agronomy: Madison, WI, USA, 1996; pp. 1201–1229.

57. ISO 11261. *Soil Quality—Determination of Total Nitrogen—Modified Kjeldahl Method*; International Organization for Standardization: Geneva, Switzerland, 1995.

58. Egner, H.; Riehm, H.; Domingo, W.R. Untersuchun-gen über die chemische Bodenanalyse als Grundlage für die Beurteilung des Nährsto_zustandes der Böden. II. Chemische Extractionsmethoden zur Phospor-und Kaliumbestimmung. *Ann. R. Agric. Coll. Swed.* **1960**, *26*, 199–215.

59. Schlichting, E.; Blume, H.P.; Stahr, K. *Bodenkundliches Praktikum. Pareys Studientexte*; Blackwell Wissenschafts: Berlin, Germany, 1995; p. 81.

60. ISO 11260 Preview. *Soil Quality—Determination of E_ective Cation Exchange Capacity and Base Saturation Level Using Barium Chloride Solution*; International Organization for Standardization: Geneva, Switzerland, 2018.

61. Dell Inc. *Dell Statistica (Data Analysis Software System), Version 13.1*; Dell Inc.: Tulsa, OK, USA, 2016.

62. Parks, D.H.; Tyson, G.W.; Hugenholtz, P.; Beiko, R.G. STAMP: Statistical analysis of taxonomic and functional profiles. *Bioinformatics* **2014**, *30*, 3123–3124. [CrossRef]

63. Krzywinski, M.I.; Schein, J.E.; Birol, I.; Connors, J.; Gascoyne, R.; Horsman, D.; Jones, S.J.; Marra, M.A. Circos: An information aesthetic for comparative genomics. *Genome Res.* **2009**, *19*, 1639–1645. [CrossRef]

64. Garbeva, P.; van Veen, J.A.; van Elsas, J.D. Microbial diversity in soil: Selection microbial populations by plant and soil type and implications for disease suppressiveness. *Annu. Rev. Phytopathol.* **2004**, *42*, 243–270. [CrossRef]

65. Köberl, M.; Müller, H.; Ramadan, E.M.; Berg, G. Desert farming benefits from microbial potential in arid soils and promotes diversity and plant health. *PLoS ONE* **2011**, *6*, e24452. [CrossRef]

66. Chaparro, J.M.; Sheflin, A.M.; Manter, D.K.; Vivanco, J.M. Manipulating the soil microbiome to increase soil health and plant fertility. *Biol. Fertil. Soils* **2012**, *48*, 489–499. [CrossRef]

67. Maron, P.A.; Sarr, A.; Kaisermann, A.; Lévêque, J.; Mathieu, O.; Guigue, J.; Karimi, B.; Bernard, L.; Dequiedt, S.; Terrat, S.; et al. High microbial diversity promotes soil ecosystem functioning. *Appl. Environ. Microbiol.* **2018**, *84*, 1–13. [CrossRef]

68. Doornbos, R.F.; van Loon, L.C.; Bakker, P.A.H. Impact of root exudates and plant defense signaling on bacterial communities in the rhizosphere. A review. *Agron. Sustain. Dev.* **2012**, *32*, 227–243. [CrossRef]

69. Verma, J.P.; Yadav, J.; Tiwari, K.N.; Jaiswal, D.K. Evaluation of plant growth promoting activities of microbial strains and their effect on growth and yield of chickpea (*Cicer arietinum* L.) in India. *Soil Biol. Biochem.* **2014**, *70*, 33–37. [CrossRef]

70. Kumar, A.; Maurya, B.R.; Raghuwanshi, R. Isolation and Characterization of PGPR and their effect on growth, yield and Nutrient content in wheat (*Triticum aestivum* L.). *Biocatal. Agril. Biotechnol.* **2014**, *3*, 121–128. [CrossRef]

71. Kumar, A.; Bahadur, I.; Maurya, B.R.; Raghuwanshi, R.; Meena, V.S.; Singh, D.K.; Dixit, J. Does a plant growth promoting rhizobacteria enhance agricultural sustainability? *J. Pure Appl. Microbio.* **2015**, *9*, 715–724.

72. Verma, J.P.; Yadav, J.; Tiwari, K.N.; Kumar, A. Effect of indigenous *Mesorhizobium* spp. And plant growth promoting rhizobacteria on yields and nutrients uptake of chickpea (*Cicer arietinum* L.) under sustainable agriculture. *Ecol. Eng.* **2013**, *51*, 282–286. [CrossRef]

73. Jalali, G.; Lakzian, A.; Astaraei, A.; Haddad-Mashadrizeh, A.; Azadvar, M.; Esfandiarpour, E. Effects of land use on the structural diversity of soil bacterial communities in Southeastern Iran. *Biosci. Biotech. Res. Asia.* **2016**, *13*, 1739–1747. [CrossRef]

74. Soliman, T.; Yang, S.Y.; Yamazaki, T.; Jenke-Kodama, H. Profiling soil microbial communities with next-generation sequencing: The influence of DNA kit selection and technician technical expertise. *PeerJ* **2017**, *5*, e4178. [CrossRef] [PubMed]

75. Mi, L.; Wang, G.; Jin, J.; Sui, Y.; Liu, J.; Liu, X. Comparison of microbial community structures in four Black soils along a climatic gradient in northeast China. *Can. J. Soil Sci.* **2012**, *92*, 543–549. [CrossRef]

76. Xu, Y.; Wang, G.; Jin, J.; Liu, J.; Zhang, Q.; Liu, X. Bacterial communities in soybean rhizosphere in response to soil type, soybean genotype, and their growth stage. *Soil Biol. Biochem.* **2009**, *41*, 919–925. [CrossRef]

77. Attwood, G.T.; Wakelin, S.A.; Leahy, S.C.; Rowe, S.; Clarke, S.; Chapman, D.F.; Muirhead, R.R.; Jacobs, J.M.E. Applications of the soil, plant and rumen microbiomes in pastoral agriculture. *Front. Nutr.* **2019**, *6*, 107. [CrossRef]

78. Bakker, M.G.; Chaparro, J.M.; Manter, D.K.; Vivanco, J.M. Impacts of bulk soil microbial community structure on rhizosphere microbiomes of *Zea mays*. *Plant Soil.* **2015**, *392*, 115. [CrossRef]

79. Pascault, N.; Ranjard, L.; Kaisermann, A.; Bachar, D.; Christen, R.; Terrat, S.; Mathieu, O.; Lévêque, J.; Mougel, C.; Henault, C.; et al. Stimulation of different functional groups of bacteria by various plant residues as a driver of soil priming effect. *Ecosystems* **2013**, *16*, 810–822. [CrossRef]

80. Huang, X.F.; Chaparro, J.M.; Reardon, K.F.; Zhang, R.; Shen, Q.; Vivanco, J.M. Rhizosphere interactions: Root exudates, microbes, and microbial communities. *Botany* **2014**, *92*, 267–275. [CrossRef]

81. Fernandes, C.C.; Kishi, L.T.; Lopes, E.M.; Omori, W.P.; Souza, J.A.M.; Alves, L.M.C.; Lemos, E.G. Bacterial communities in mining soils and surrounding areas under regeneration process in a former ore mine. *Braz. J. Microbiol.* **2018**, *49*, 489–502. [CrossRef] [PubMed]

82. Nelkner, J.; Henke, C.; Lin, T.W.; Pätzold, W.; Hassa, J.; Jaenicke, S.; Grosch, R.; Puhler, A.; Sczyrba, A.; Schlüter, A. Effect of Long-Term Farming Practices on Agricultural Soil Microbiome Members Represented by Metagenomically Assembled Genomes (MAGs) and Their Predicted Plant-Beneficial Genes. *Genes* **2019**, *10*, 424. [CrossRef]

83. Fei, Y.; Huang, S.; Zhang, H.; Tong, Y.; Wen, D.; Xia, X.; Wang, H.; Luo, Y.; Barceló, D. Response of soil enzyme activities and bacterial communities to the accumulation of microplastics in an acid cropped soil. *Sci. Total Environ.* **2019**, 135634. [CrossRef]

84. Yeager, C.M.; Gallegos-Graves, L.V.; Dunbar, J.; Hesse, C.N.; Daligault, H.; Kuske, C.R. Polysaccharide Degradation Capability of Actinomycetales Soil Isolates from a Semiarid Grassland of the Colorado Plateau. *Appl. Environ. Microbiol.* **2017**, *83*, 6. [CrossRef]

85. Jog, R.; Nareshkumar, G.; Rajkumar, S. Plant growth promoting potential and soil enzyme production of the most abundant *Streptomyces* pp. from wheat rhizosphere. *J. Appl. Microbiol.* **2012**, *113*, 1154–1164. [CrossRef]

86. Lopes, A.A.C.; de Sousa, D.M.G.; Chaer, G.M.; Junior, F.B.R.; Goedert, W.J.; Mendes, I.C. Interpretation of microbial soil indicators as a function of crop yield and organic carbon. *Soil Sci. Soc. Am. J.* **2013**, *77*, 461–472. [CrossRef]

87. Stott, D.E.; Cambardella, C.A.; Tomer, M.D.; Karlen, D.L.; Wolf, R. A soil quality assessment within the Iowa River South Fork watershed. *Soil Sci. Soc. Am. J.* **2011**, *75*, 2271–2282. [CrossRef]

88. Borowik, A.; Wyszkowska, J.; Kucharski, M.; Kucharski, J. Implications of soil pollution with diesel oil and BP Petroleum with Active Technology for soil health. *Int. J. Environ. Res. Public Health* **2019**, *16*, 2474. [CrossRef] [PubMed]

89. Kumar, S.; Chaudhuri, S.; Maiti, S.K. Soil dehydrogenase enzyme activity in natural and mine soil—A Review. *Middle-East J. Sci. Res.* **2013**, *13*, 898–906. [CrossRef]

90. Merino, C.; Godoy, R.; Matus, F. Soil enzymes and biological activity at different levels of organic matter stability. *J. Soil Sci. Plant Nutr.* **2016**, *16*, 14–30. [CrossRef]

91. Bautista-Cruz, A.; Ortiz-Hernández, Y.D. Hydrolytic soil enzymes and their response to fertilization: A short review. *Commun. Sci.* **2015**, *6*, 255–262. [CrossRef]

92. Foster, E.J.; Fogle, E.J.; Cotrufo, M.F. Sorption to biochar impacts β-glucosidase and phosphatase enzyme activities. *Agriculture* **2018**, *8*, 158. [CrossRef]

Article

Nitrogen Fixing and Phosphate Mineralizing Bacterial Communities in Sweet Potato Rhizosphere Show a Genotype-Dependent Distribution

Joana Montezano Marques [1], Jackeline Rossetti Mateus [2], Thais Freitas da Silva [2], Camila Rattes de Almeida Couto [2], Arie Fitzgerald Blank [3] and Lucy Seldin [2,*

[1]   Instituto de Ciências Biológicas, Centro de Genômica e Biologia de Sistemas, Universidade Federal do Pará, Av. Augusto Corrêa, 01, Guamá, CEP 66.075-110, Belém, PA, Brazil; jomontezanomarques@gmail.com
[2]   Instituto de Microbiologia Paulo de Góes (IMPPG), Universidade Federal do Rio de Janeiro, Centro de Ciências da Saúde, Bloco I, Ilha do Fundão, CEP 21941-590, Rio de Janeiro, RJ, Brazil; jacky.rossetti@gmail.com (J.R.M.); thaisfs.bio@gmail.com (T.F.d.S.); camilarattes@gmail.com (C.R.d.A.C.)
[3]   Departamento de Engenharia Agronômica, Universidade Federal de Sergipe, Av. Marechal Rondon S/N, CEP 49100-000, São Cristóvão, SE, Brazil; arie.blank@gmail.com
*   Correspondence: lseldin@micro.ufrj.br; Tel.: +55-213-938-6741; Fax: +55-212-560-8344

Received: 23 October 2019; Accepted: 28 November 2019; Published: 3 December 2019

Abstract: We hypothesize that sweet potato genotypes can influence the bacterial communities related to phosphate mineralization and nitrogen fixation in the rhizosphere. Tuberous roots of field-grown sweet potato from genotypes IPB-149, IPB-052, and IPB-137 were sampled three and six months after planting. The total community DNA was extracted from the rhizosphere and analyzed by Polymerase Chain Reaction-Denaturing Gradient Gel Electrophoresis (PCR-DGGE) and quantitative real-time PCR (qPCR), based on the alkaline phosphatase coding gene (*alp* gene) and on the nitrogenase coding gene (*nifH* gene). The cluster analysis based on DGGE showed that plant age slightly influenced the bacterial community related to phosphate mineralization in the rhizosphere of IPB-137, although it did not affect the bacterial community related to nitrogen fixation. The statistical analysis of DGGE fingerprints (Permutation test, $p \leq 0.05$) showed that nitrogen-fixing bacterial community of IPB-052 statistically differed from genotypes IPB-149 and IPB-137 after six months of planting. The bacterial community of IPB-137 rhizosphere analyzed by *alp* gene also showed significant differences when compared to IPB-149 in both sampling times ($p \leq 0.05$). In addition, *alp* gene copy numbers significantly increased in abundance in the rhizosphere of IPB-137 after six months of planting. Therefore, plant genotype should be considered in the biofertilization of sweet potato.

Keywords: sweet potato; bacterial communities; nitrogen fixation; phosphate mineralization; plant genotype

---

1. Introduction

The rhizosphere is characterized as the soil in close contact with plant roots and is where the root microbiome recruitment occurs through exudation of plant molecular signals, especially secondary metabolites [1,2]. Plants are usually benefited by the interactions with soil microbes in their rhizospheres, improving plant nutrient acquisition, pathogen resistance, and stress tolerance [3,4]. The so-called 'Plant Growth-Promoting Rhizobacteria' (PGPR) are non-pathogenic bacteria that live in the rhizosphere soil and are able to promote plant development [3,5–7]. Some PGPR are able to provide essential nutrients to plants using a compound that is synthesized by the bacterium or making some macro and micro-nutrients available that were immobilized in mineral and organic compounds, and were not free for plant uptake [8]. The biological nitrogen fixation and the phosphate solubilization/mineralization are examples of the direct promotion of plant growth by PGPR [9]. The use of PGPR in biofertilization is an open field for research on sustainable agriculture, as it optimizes crop yields based on beneficial plant-microbe interactions [1,10]. The use of PGPR has been tested with success in different plants such as maize, potato, wheat, and many other economically important ones [11].

A large number of biotic and abiotic factors influence the microbial communities in the rhizosphere. Physicochemical characteristics of the soil directly modulate microbial communities [2] and plant physiology and genetics also control rhizosphere composition [1,12]. The impact of plant genotypes on rhizosphere microbiome composition varies depending on soil context and plant species studied [13]. Therefore, understanding mutual adaptation between microbes and plants in response to different environmental conditions can contribute to crop breeding and management programs.

Sweet potato plants (*Ipomoea batatas* L.) present a complex root system composed by fibrous roots - specialized in nutrient absorption - and tuberous roots - specialized in nutrient storage. The transition from fibrous root to tuberous root is related to starch accumulation [14,15]. The tuberous roots represent one of the major nutrient sources for countries in development, and they also contribute to a supply of vitamins B, C, and E and minerals, including iron, calcium, zinc, and selenium, in the human diet (International Potato Center – [16]). Sweet potato is cultivated worldwide, with Asia and Africa together producing 95% of all roots commercialized globally [17]. Sweet potato is used for food, animal feed, and processing (as food, starch, and other products) and its biggest global producer and consumer is China. Among Latin America countries, Brazil is in the spotlight as the main sweet potato producer. Sweet potato is considered to be the country's fourth most consumed vegetable crop, and in the northeast of the country, the culture is important both economically and socially [18].

For sweet potato, as for other crops, high root yields are desirable for good crop productivity. Usually, chemical fertilizers are used, but the high inorganic fertilizer input contributes negatively to production costs and to environment pollution [19]. They are expensive, non-eco-friendly, cause eutrophication, reduce organic matter and microbial activity in soil, and are hazardous to health [20]. Therefore, there is increasing interest in the use of biofertilizers for sustainable agriculture. PGPR can help to increase quality of the soil by providing nutrients required for benefit of the plants [8]. Dawwam, et al. [21] have already demonstrated the beneficial effect of PGPR isolated from the roots of sweet potato. Different isolates having abilities for IAA production and phosphate solubilization were tested as bioinoculants to potato tubers, and the inoculated plants showed significant differences in various parameters such as vegetative growth, photosynthetic pigments, and N, P, and K concentrations.

According to other studies, the growth of sweet potato can be significantly improved by PGPR [19,22–24]. However, information about the PGPR communities during the sweet potato growth and their behavior depending on the genotype of sweet potato studied is still limited. In previous studies, the total bacterial communities from the rhizosphere and endosphere of the tuberous roots of three sweet potato genotypes (IPB-149, IPB-137, and IPB-052) were characterized by Marques, et al. [12,25]. The results showed a strong rhizosphere effect in the soil surrounding the sweet potato tuberous roots, showing an influence of plant age and genotypes in bacterial communities. However, the effects of either plant age or genotypes specifically in the PGPR communities (nitrogen fixing and phosphate mineralizing bacteria) found in the rhizosphere of these sweet potato genotypes have never been described before. Therefore, in the present study, we used cultivation-independent methods (PCR-DGGE and quantitative PCR) based on *alp* gene (alkaline phosphatase coding gene) and on *nifH* gene (nitrogenase coding gene) to analyze the structure and the density (gene copy numbers) of the bacterial communities related to biofertilization (phosphate mineralization and nitrogen fixation) in rhizosphere soil. The analyses were performed to compare the distribution of *alp* and *nifH* genes among the sweet potato genotypes sampled three and six months after planting. The data generated in this study could be important for increasing knowledge on the productivity of these sweet potato genotypes under field conditions.

2. Materials and Methods

2.1. Plant Genotypes, Experimental Field Design and Conditions

The sweet potato genotypes IPB-149, IPB-137, and IPB-052 used in the present study are from the Active Germplasm Bank of the Federal University of Sergipe (UFS). These genotypes were previously described by Marques, et al. [12,25]. In brief, the IPB-149 genotype, the most commercialized genotype in the northeast of Brazil, presents a white surface color in the tuberous roots, a high starch content, and a major resistance to insect attack. The IPB-052 genotype, commercially known as cultivar Brazlândia Rosada, shows similar characteristics to the IPB-149 genotype. In contrast, the IPB-137 genotype, from the breeding program of the Federal University of Lavras, Brazil and denoted as clone 2007HSF028-08, presents different characteristics, including a pink color on its tuberous root surface, a lower starch content, and a low resistance to insect attack. It has longer branches and a shorter length between nodes, and the use of its aerial part for animal feed purposes has been suggested. Other morpho-agronomic traits, including the dry mass of aboveground part and the total productivity of the roots, were presented in Alves, et al. [26].

In 2011, a field experiment with sweet potato plants was performed in the Research Farm "Campus Rural da UFS". The experimental farm is located in the 'São Cristóvão' municipality (10°55′27″ S/ 37°12′01″ W), Sergipe State, northeast of Brazil. The experimental farm soil characteristics were: pH (5.4), Ca^{2+} (0.82 cmolc/dm^3), Mg^{2+} (0.43 cmolc/dm^3), Al^{3+} (0.65 cmolc/dm^3), Na (3.5 mg/dm^3), K (21.1 mg/dm^3), P (7.0 mg/dm^3), 73.82% sand, 20.72% silt, 5.46% clay, and 0.86% C_{org} [14–16]. The experimental field design was described in Marques, et al. [12,25]. The genotypes were planted in three replicate rows (randomized block design) in the experimental plot with spaces of 0.8 m between the rows and 0.35 m between plants. All of the cultural management processes used were those described in Alves, et al. [26]. After three (t1, _3M) and six (t2, _6M) months, sweet potato plants with similar plant growth developmental stages were randomly harvested from the research farm soil. A total of five plants/replicates of the three different sweet potato genotypes were collected in each sampling time. Their roots were shaken to remove the loosely attached soil, and the soil still adhering to the tuberous roots was aseptically brushed off from all tuberous roots of every plant. This soil was considered to be rhizosphere soil. The rhizosphere samples were kept at −20 °C before total community DNA (TC-DNA) extraction.

2.2. TC-DNA Extraction and PCR Amplification

The TC-DNA was extracted from each five replicates of rhizosphere (0.5 g of each) from the three sweet potato genotypes (IPB-149, IPB-137, and IPB-052), and from both sampling times (t1 and t2). The Fast DNA Spin Kit for soil (Qbiogene, BIO 101 Systems, Carlsbad, CA, USA) was used according to the manufacturer's instructions. TC-DNA preparations were used in PCR reactions in order to amplify the *alp* gene (alkaline phosphatase coding gene) and *nifH* gene (nitrogenase coding gene).

The PCR conditions for *alp* gene amplification were performed as described by Sakurai, et al. [27]. The primers ALPS-F730/ALPS-R1101 were used in the PCR reactions and the reverse primer was added with a GC clamp. The reaction conditions were those previously described in Sakurai, et al. [27]. Fragments of the *nifH* gene were amplified using a nested-PCR approach. Briefly, the primers FGPH19 [28] and PolR [29] were used for the first-round of the PCR. The first-round PCR conditions were those described by Monteiro, et al. [30]. A 1:100 dilution of the first-round PCR product was used as a template for the second-round with the primers PolF/AQER using the same reaction conditions as described by Poly, et al. [29]. The PCR products were visualized by electrophoresis and then stored at −20 °C until DGGE analyses.

2.3. DGGE and Statistical Analyses

The INGENYphorU-2 system (INGENY International BV, Middelburg, The Netherlands) was used for DGGE. The PCR products from both genes (*alp* and *nifH*) and a standard bacterial marker (previously described by Heuer, et al. [31]) were loaded directly to DGGE. The denaturing gradient of urea and formamide varied in the range 40%–60% (the *alp* gene), and 40%–70% (the *nifH* gene). The electrophoresis conditions were performed as described by Sakurai, et al. [27] and Monteiro, et al. [30] for the *alp* and the *nifH* genes, respectively. After electrophoresis, the DGGEs were stained with SYBR Green I and visualized using a STORM apparatus (Amersham Pharmacia Biotech, Munich, Germany). The unweighted pair group with average linkages (UPGMA) was used for cluster analysis. The software BioNumerics 5.0 version (Applied Mathematics, Kortrijk, Belgium) was used to construct dendrograms based on Pearson similarity indices. Significant differences ($p \leq 0.05$) between DGGE profiles were determined by permutation tests based on pairwise Pearson correlation indices using the PERMTEST software [32]. In addition, quantitative matrices generated from the DGGE lanes based on Dice correlation index were exported to PAST software [33] for principal component analysis (PCA).

2.4. Determination of alp *and* nifH *Genes Copy Numbers*

The abundance of phosphate mineralizing and nitrogen fixing bacteria were quantified by targeting the *alp* and *nifH* genes, respectively, in a quantitative real-time PCR (qPCR). The reactions were performed in an ABI Prism 7300 Cycler (Applied Biosystems, Germany) in 25 µL reaction mixtures containing $1 \times$ SYBR^{TM} Select Master Mix (Applied Biosystems, Germany), 0.4 µg/µL BSA, 1 µL of target DNA (approximately 50 ng) and 0.5 µM of primers. For the quantification of *alp* gene, primers ALPS-F730/ALPS-R1101 [27] were used. The amplification conditions were 2 min at 50 °C, 10 min at 95 °C, and 40 cycles consisting of 30 s at 95 °C, 60 s at 57 °C, and 30 s at 72 °C. Standard curves were prepared by serial dilutions (from 10^8 to 10^1 gene copy numbers/µL) of plasmid DNA (pTZ57R/T; CloneJET^{TM} PCR Cloning Kit - Thermo Fisher Scientific, Waltham, MA, USA) containing the cloned *alp* gene PCR products from bacterial strain 006.30 [34]. For the quantification of the *nifH* gene, primers FGPH19 [28] and PolR [29] were chosen as described by Taketani, et al. [35]. The amplification conditions were an initial denaturation step at 94 °C for 10 min, followed by 40 cycles of 1 min at 94 °C, 27 s at 57 °C, and 1 min at 72 °C. Standard curves were obtained using serial dilutions of the *Escherichia coli*-derived vector plasmid JM109 (Promega, Madison, WI, USA) containing a cloned *nifH* gene from *Bradyrhizobium liaoningense*, using 10^2 to 10^7 gene copies/µL. Significant differences between samples were tested in pairwise comparisons using the Tukey test ($p \leq 0.05$; SAS 9.3; SAS Institute Inc., Cary, NC, USA).

3. Results

3.1. Structure of Phosphate Mineralizing and Nitrogen Fixing Bacterial Communities Analyzed by DGGE

The effects of plant growth stage and genotypes on the structure of the phosphate mineralizing and nitrogen fixing bacterial communities present in the rhizosphere of tuberous roots of the genotypes IPB-149, IPB-137, and IPB-052 were analyzed by DGGE based on *alp* gene and *nifH* gene fragments, respectively, which were amplified from TC-DNA.

For the DGGE profiles based on the *alp* gene, principal component analysis (PCA) showed the formation of two separate groups: IPB-137_6M and IPB-149_3M (Figure 1A). The fingerprints of the phosphate mineralizing bacterial community in the rhizosphere were statistically analyzed using the permutation test ($p \leq 0.05$). The sampling time had a significant effect on the structure of the bacterial community in the IPB-137 genotype (Table 1). Moreover, the results showed that the genotypes also significantly influenced the structure of the phosphate mineralizing bacterial community in IPB-137 and IPB-149 (Figure 1B). These communities were statistically different for both sampling times (Table 1). Therefore, the different genotypes and, to a minor extent, the time influenced the phosphate mineralizing bacterial community present in the tuberous roots of sweet potato studied here. The DGGE fingerprint analyses showed that the structure of both bacterial communities were complex with high variability among the replicates (Figure S1A,B). The samplings of IPB-137 and of

IPB-149 were grouped separately (per genotype) with more than 60% similarity after UPGMA cluster analysis (Figure S1A). A high similarity (more than 80%) was observed within IPB-137_6M samples, suggesting a slight influence of the plant age, as was observed in PCA.

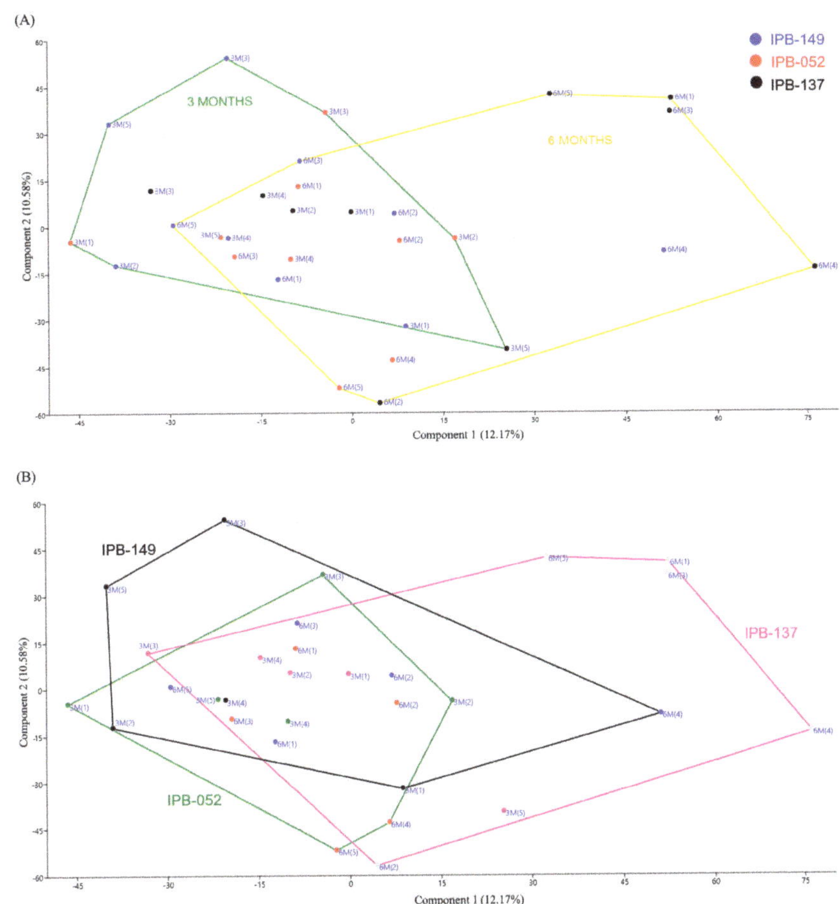

Figure 1. Principal component analyses (PCA) was conducted using Denaturing Gradient Gel Electrophoresis (DGGE) patterns of phosphate mineralizing bacteria based on the *alp* gene from the rhizosphere of three different sweet potato genotypes (IPB-149, IPB-137, and IPB-052) sampled after three and six months after planting (t1, _3M and t2, _6M, respectively). (**A**) and (**B**) highlight the grouping observed within the sampling time (t1 × t2) and within the different genotypes, respectively.

For the DGGE profiles based on *nif* gene, the influence of the sampling time was not evident in PCA, except for some of IPB-137_6M replicates (Figure 2A). The results also showed that the genotypes affected the structure of the nitrogen fixing bacterial community (Figure 2B). The fingerprints of the nitrogen fixing bacterial community in the rhizosphere were statistically analyzed using the permutation test ($p \leq 0.05$). No significant effect of the sampling time on the structure of the nitrogen fixing bacterial community within the different genotypes was observed (Table 1). In contrast, the structure of the nitrogen fixing bacterial communities in sweet potato genotypes IPB-149 and IPB-137 were statistically different from those in IPB-052 after six months of planting (Table 1). The visual interpretation of the DGGE profiles corroborated the PCA. UPGMA cluster analysis of the DGGE profiles based on *nifH* gene showed a high similarity within the replicates of IPB-137_6M. The replicates of IPB-052_6M formed a group that was separate from the other genotypes (Figure S1B).

Table 1. Dissimilarity (d value in %) of rhizosphere bacterial fingerprints (DGGE) based on *alp* and *nifH* genes with comparisons between samplings (t1, _3M and t2, _6M) or among IPB-149, IPB-137, and IPB-052 genotypes.

Genes	Sampling Time (t1 × t2)	Dissimilarity (%)	Genotypes (t1)	Dissimilarity (%)	Genotypes (t2)	Dissimilarity (%)
alp	IPB-149	0.7	IPB-149 × IPB-137	8.1*	IPB-149 × IPB-137	17.6*
	IPB-137	0.4*	IPB-149 × IPB-052	0.2	IPB-149 × IPB-052	3.1
	IPB-052	−0.3	IPB-137 × IPB-052	−1.2	IPB-137 × IPB-052	4.1
nifH	IPB-149	−1.9	IPB-149 × IPB-137	−1.8	IPB-149 × IPB-137	9.0
	IPB-137	18.3	IPB-149 × IPB-052	0.4	IPB-149 × IPB-052	11.0*
	IPB-052	12.4	IPB-137 × IPB-052	1.6	IPB-137 × IPB-052	26.8*

* Significant difference ($p \leq 0.05$) as determined using the permutation test.

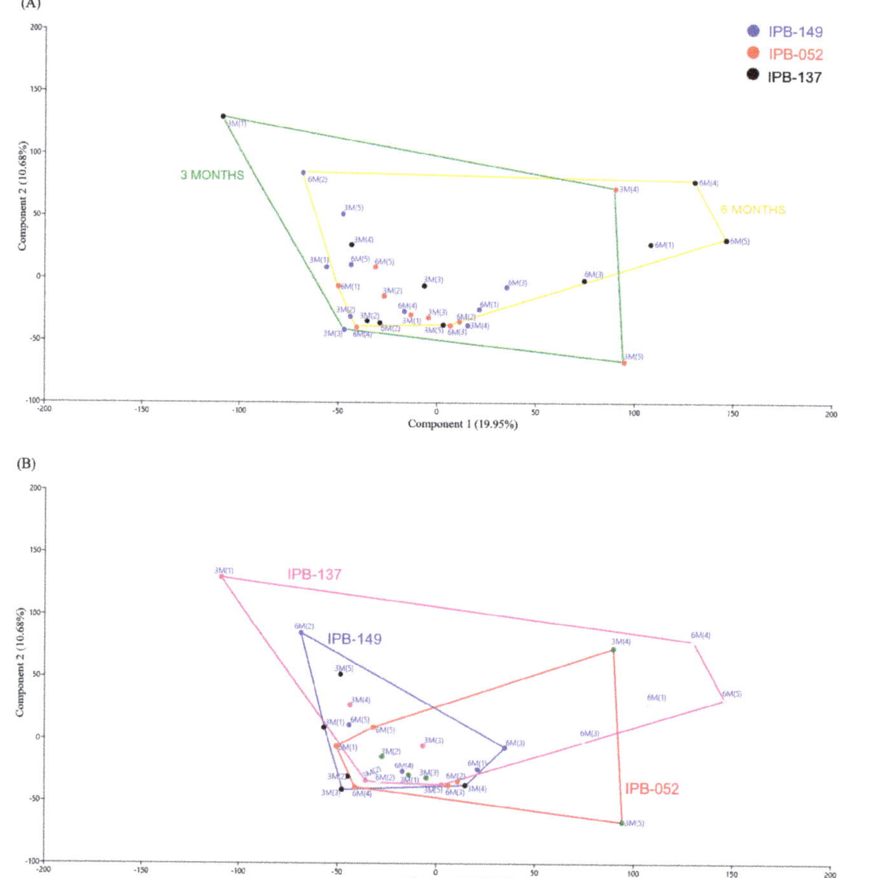

Figure 2. Principal component analyses (PCA) were conducted using DGGE patterns of nitrogen fixing bacteria based on the *nifH* gene from the rhizosphere of three different sweet potato genotypes (IPB-149, IPB-137, and IPB-052) sampled after three and six months after planting (t1, _3M and t2, _6M, respectively). (A) and (B) highlight the grouping observed within the sampling time (t1 x t2) and within the different genotypes, respectively.

3.2. Quantitative Real-time PCR of alp and nifH Genes

The *alp* and *nifH* genes were quantified in all sweet potato rhizospheres. The values for *alp* gene quantification (copies of *alp* gene/g of rhizosphere soil) were higher than for *nifH* gene (Figure 3). The analyses revealed a statistically significant difference ($p \leq 0.05$) in the abundance levels of the *alp* genes only in IPB-137 after six months of planting, suggesting that the abundance of the phosphate mineralizing bacterial community present in the tuberous roots of sweet potato is influenced by the genotype and also by the time in IPB-137 (Figure 3A). Similarly, the IPB-137 genotype showed statistically significant differences in the abundance levels of the *nifH* genes between three and six months (t1 × t2) after planting (Figure 3B). Statistically similar *nifH* gene densities ($p \leq 0.05$) were observed between the t1 and t2 samplings in IPB-149 and IPB-052. When comparing the t2 samplings, no differences in *nifH* gene densities were observed between the different genotypes (Figure 3B).

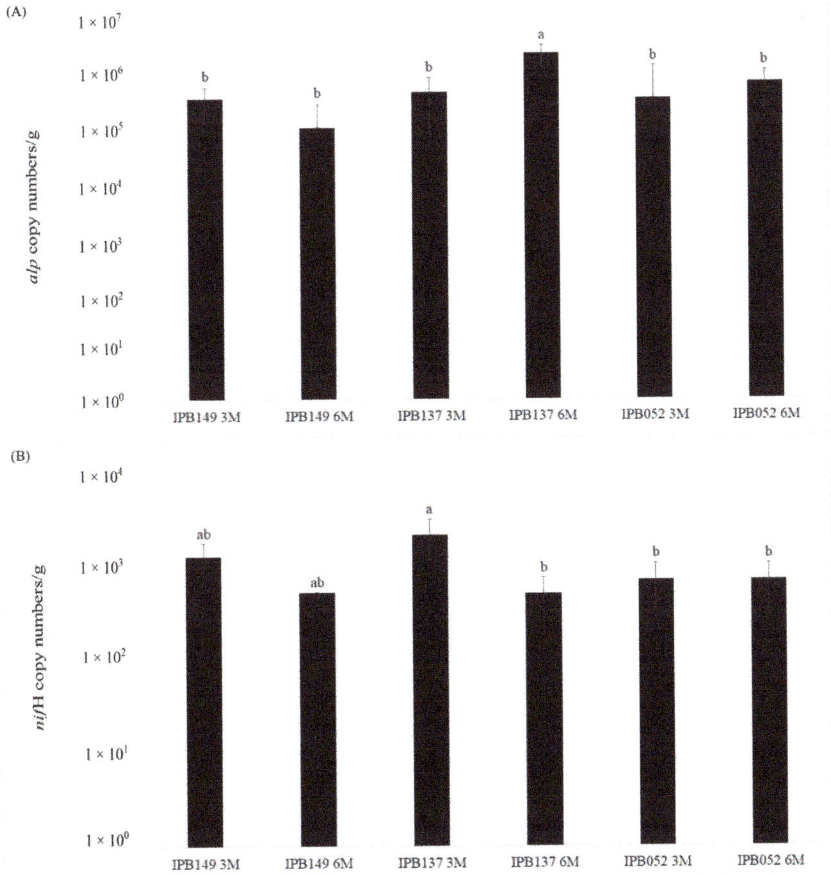

Figure 3. Abundance of bacterial phosphate mineralizing bacteria based on the *alp* gene (**A**) and nitrogen fixing bacteria based on the *nifH* gene (**B**) in the rhizosphere of three different sweet potato genotypes (IPB-149, IPB-137 and IPB-052) sampled after three and six months after planting (t1, _3M and t2, _6M, respectively). The error bars indicate standard error. Different letters (a and b) indicate significant differences of means in pairwise comparisons (Tukey test; $p \leq 0.05$).

4. Discussion

The use of biofertilizers, including plant growth promoting bacteria, is increasing worldwide as an environmentally-friendly alternative for soil fertility management, sustaining not only sweet potato yields (both in terms of tuber weight and in terms of the levels of starch in sweet potato crops) but also different economically important plants [11], compared with the continuous application of inorganic fertilizers alone [36–39]. The plant positive response to biofertilization stimulates the continuous optimization of crop yields by exploiting beneficial plant-microbe interactions [4]. However, the effect of different biofertilizers on root colonization, nutrient uptake, growth, and yield is not always as effective as expected. Different types of soils found in various countries, as well as plant genotypes and the sampling time (period the crop was maintained in the field) all contribute to different responses in plant yields, including those observed in sweet potato [22,36,37].

It was evident in previous studies conducted by our group that the total bacterial communities associated with different genotypes of sweet potato planted in Brazil varied depending on the plant age and genotype [12,25]. However, the behavior of specific communities such as phosphate mineralizers and nitrogen fixers, which may contribute directly to the plant development, was not taken into account at that time. Indeed, the majority of the studies focusing on phosphate-solubilizing/mineralizing and nitrogen-fixing bacterial communities found in the literature were conducted to screen for beneficial traits that could improve plant growth. Alterations on the structure and abundance of these communities along the plant growth have not been accessed, which are usually restricted to a single genotype and a growth stage of the plant [21,24,40].

In the present study, we demonstrated that the structure of phosphate- mineralizing bacterial community (based on the presence of *alp* genes) varied along the plant growth depending on the genotype considered. The structure of the phosphate mineralizing bacterial community in IPB-137 was mainly different from that in IPB-149. Tarafdar and Jungk [41] observed that both acid and alkaline phosphatase activities of soil were usually increased near the rhizoplane of different plants and such an increase depended on plant species, soil type, and plant age. Although the optimal pH for the activity of the alkaline phosphatase is higher than the pH of the soil where the sweet potatoes were planted, its role remains an important one regardless. For example, both acid and alkaline phosphatase activities near the rhizoplane of maize resulted in alterations of the composition of bacterial communities, as determined by PCR-DGGE [42]. In addition to the influence of the genotype, a statistically difference in phosphate mineralizing bacterial communities in IPB-137 was observed between samplings after three to six months of planting. Changes were observed not only in the structure of this community but also in the abundance of *alp* genes after six months of growth. These changes could be explained by the fact that, at the late growth stage of sweet potato, the roots usually presented stronger nutrient absorption ability due to the increase of the number of root tips, the enlargement of root surface area, and the root volume [43]. Phosphate mineralizers may have contributed to the conversion of insoluble phosphates into available forms for plant via different processes such as acidification, exchange reactions, chelation, and production of gluconic acid [21,44].

Nitrogen-fixing bacteria identified as *Azospirillum* sp. were first isolated from the fibrous roots and storage root peels of sweet potato [45]. From this time on, different studies were carried out on the isolation and identification of nitrogen fixers in sweet potato roots, as important evidence indicated that nonsymbiotic N_2 fixation in sweet potato is agronomically significant [40]. Active expressions of the *nifH* gene phylogenetically similar to those of *Bradyrhizobium* spp., *Sinorhizobium* sp., *Azorhizobium* sp., *Bacillus* sp. and *Pelomonas* sp. were found in large quantities in the N_2-fixing sweet potato storage tubers planted in different regions [40,46,47]. In our study, the structure and abundance of nitrogen-fixing bacterial communities based on the *nifH* gene were determined and compared between the three genotypes.

The nitrogen-fixing bacterial community in the rhizosphere of IPB-052 seems to be different from those of the other two genotypes. IPB-052 presents a white surface color on the tuberous roots and high starch content. It is quite possible that specific nitrogen-fixing bacteria are being recruited from soil to the rhizosphere. This rhizosphere effect can generate a zone with maximum microbial activity, resulting in a pool of metabolic products that can support plant growth [48]. However, the difference observed on the structure of the nitrogen-fixing bacterial community in IPB-052 does not also appear in the analysis of the *nifH* gene abundance. The *nifH* gene copy numbers appeared to be lower than those of *alp* genes, and the only difference observed among the genotypes studied was between the 3–6 month samplings in IPB-137. This fact suggests that nitrogen-fixing bacteria may only contribute more towards plant development during the early stages of plant growth in this sweet potato genotype.

5. Conclusions

The results obtained here suggest that the three sweet potato genotypes studied showed a differentiated recruitment of members of the studied communities (phosphate mineralizers and nitrogen fixers). In addition, the IPB-137 genotype showed the highest number of copies of both genes studied with statistically significant differences compared to the other genotypes. Even though we know that the distribution of these communities may be directly related to the productivity of a given genotype, more studies are needed to trace this correlation in the field.

Supplementary Materials: The following are available online at http://www.mdpi.com/1424-2818/11/12/231/s1, Figure S1. DGGE fingerprints and UPGMA cluster analyses based on *alp* gene fragments (A) and *nifH* gene fragments (B) of phosphate mineralizing and nitrogen fixing bacterial communities, respectively, present in the rhizosphere of the tuberous roots (replicates 1–5 in parenthesis) from three different sweet potato genotypes (IPB-149, IPB-137 and IPB-052) sampled three (t1, _3M - black squares) and six (t2, _6M - grey squares) months after planting.

Author Contributions: Conceptualization, L.S. and J.M.M.; Data curation, A.F.B. and J.M.M.; Formal analysis, C.R.d.A.C., J.M.M., J.R.M., T.F.d.S. and L.S.; Funding acquisition, L.S.; Methodology, C.R.d.A.C., J.M.M., J.R.M. and T.F.d.S.; Project administration, L.S.; Resources, A.F.B.; Supervision, L.S.; Writing—original draft, J.M.M. and L.S.; Writing—review & editing, A.F.B., C.R.d.A.C., J.M.M., J.R.M., T.F.d.S. and L.S.

Funding: Coordenação de Aperfeiçoamento de Pessoal de Nível Superior: 01; Conselho Nacional de Desenvolvimento Científico e Tecnológico; Fundação Carlos Chagas Filho de Amparo à Pesquisa do Estado do Rio de Janeiro.

Acknowledgments: The authors thank Simone Cotta for her help in qPCR experiments. This study was financially supported by the National Research Council of Brazil (CNPq), Fundação de Amparo à Pesquisa do Estado do Rio de Janeiro (FAPERJ) and Coordenação de Aperfeiçoamento de Pessoal de Nível Superior (CAPES).

Conflicts of Interest: The authors declare that they have no conflict of interest.

References

1. Berendsen, R.L.; Pieterse, C.M.J.; Bakker, P.A.H.M. The rhizosphere microbiome and plant health. *Trends Plant Sci.* **2012**, *17*, 478–486. [CrossRef] [PubMed]

2. Philippot, L.; Raaijmakers, J.M.; Lemanceau, P.; van der Putten, W.H. Going back to the roots: The microbial ecology of the rhizosphere. *Nat. Rev. Microbiol.* **2013**, *11*, 789–799. [CrossRef] [PubMed]

3. Dutta, S.; Podile, A.R. Plant growth promoting rhizobacteria (PGPR): The bugs to debug the root zone. *Crit. Ver. Microbiol.* **2010**, *36*, 232–344. [CrossRef]

4. Liu, F.; Hewezi, T.; Lebeis, S.L.; Pantalone, V.; Grewal, O.S.; Staton, M.E. Soil indigenous microbiome and plant genotypes cooperatively modify soybean rhizosphere microbiome assembly. *BMC Microbiol.* **2019**, *19*, 201. [CrossRef] [PubMed]

5. Babalola, O.O. Beneficial bacteria of agricultural importance. *Biotechnol. Lett.* **2010**, *32*, 1559–1570. [CrossRef]

6. Kloepper, J.W.; Lifshitz, R.; Zablotowicz, R.M. Free-living bacterial inocula for enhancing crop productity. *Trends Biotechnol.* **1989**, *7*, 39–43. [CrossRef]

7. Vejan, P.; Abdullah, R.; Khadiran, T.; Ismail, S.; Nasrulhaq Boyce, A. Role of Plant Growth Promoting Rhizobacteria in Agricultural Sustainability—A Review. *Molecules* **2016**, *21*, 573. [CrossRef]

8. Egamberdieva, D.; Lugtenberg, B. Use of plant growth-promoting rhizobacteria to alleviate salinity stress in plants. In *Use of Microbes for the Alleviation of Soil Stresses*; Miransari, M., Ed.; Springer: New York, NY, USA, 2014; pp. 73–96.

9. Glick, B.R. The enhancement of plant growth by free-living bacteria. *Can. J. Microbiol.* **1995**, *41*, 109–117. [CrossRef]

10. Schlaeppi, K.; Bulgarelli, D. The plant microbiome at work. *Mol. Plant Microbe Interact.* **2015**, *28*, 212–217. [CrossRef]

11. Souza, R.; Ambrosini, A.; Passaglia, L.M.P. Plant growth-promoting bacteria as inoculants in agricultural soils. *Genet. Mol. Biol.* **2015**, *38*, 401–419. [CrossRef]

12. Marques, J.M.; da Silva, T.F.; Vollú, R.E.; Blank, A.F.; Ding, G.-C.; Seldin, L.; Smalla, K. Plant age and genotype affect the bacterial community composition in the tuber rhizosphere of field-grown sweet potato plants. *FEMS Microbiol. Ecol.* **2014**, *88*, 424–435. [CrossRef] [PubMed]

13. Pérez-Jaramillo, J.E.; Mendes, R.; Raaijmakers, J.M. Impact of plant domestication on rhizosphere microbiome assembly and functions. *Plant Mol. Biol.* **2016**, *90*, 635–644. [CrossRef] [PubMed]

14. Hennion, N.; Durandb, M.; Vrieta, C.; Doidya, J.; Maurousseta, L.; Lemoinea, R.; Pourtaua, N. Sugars en route to the roots. Transport; metabolism and storage within plant roots and towards microorganisms of the rhizosphere. *Physiol. Plant.* **2019**, *165*, 44–57. [CrossRef] [PubMed]

15. Li, X.Q.; Zhang, D. Gene expression activity and pathway selection for sucrose metabolism in developing storage root of sweet potato. *Plant Cell Physiol.* **2003**, *44*, 630–636. [CrossRef]

16. CIP. The International Potato Center. 2019. Available online: https://cipotato.org (accessed on 29 April 2019).

17. El Sheikha, A.F.; Ray, R.C. Potential impacts of bioprocessing of sweet potato: Review. *Crit. Rev. Food Sci. Nutr.* **2017**, *57*, 455–471. [CrossRef]

18. Albuquerque, J.R.T.; Ribeiro, R.M.P.; Pereira, L.A.F.; Barros Júnior, A.P.; da Silveira, L.M.; dos Santos, M.G.; de Souza, A.R.E.; Lins, H.A.; Bezerra Neto, F. Sweet potato cultivars grown and harvested at different times in semiarid Brazil. *Afr. J. Agric. Res.* **2016**, *11*, 4810–4818.

19. Yasmin, F.; Othman, R.; Sijam, K.; Saad, M.S. Effect of PGPR inoculation on growth and yield of sweet potato. *J. Biol. Sci.* **2007**, *7*, 421–424.

20. Sahoo, R.K.; Bhardwaj, D.; Tuteja, N. Biofertilizers: A Sustainable Eco-Friendly Agricultural Approach to Crop Improvement. In *Plant Acclimation to Environmental Stress*; Tuteja, N., Gill, S.S., Eds.; Springer: New York, NY, USA, 2013; pp. 403–432.

21. Dawwam, G.E.; Elbeltagy, A.; Emara, H.M.; Abbas, I.H.; Hassan, M.M. Beneficial effect of plant growth promoting bacteria isolated from the roots of potato plant. *Ann. Agric. Sci.* **2013**, *58*, 195–201. [CrossRef]

22. Nasution, R.A.; Tangapo, A.; Taufik, I.; Aditiawati, P. Comparison of plant growth promoting rhizobacteria (PGPR) diversity and dynamics during growth of Cilembu sweet potato (*Ipomoea batatas* L var. Rancing) in Cilembu and Jatinangor site, Indonesia. *J. Pure Appl. Microbiol.* **2017**, *11*, 837–845. [CrossRef]

23. Radziah, O.; Zulkifli, H.S. Utilization of Rhizobacteria for Increased Growth of Sweet Potato. In *Investing Innovation: Agriculture; Food and Forestry*; Zaharah, A.R., Ed.; University Putra Malaysia Press: Serdang, Malaysia, 2003; pp. 255–258.

24. Yasmin, F.; Othman, R.; Sijam, K.; Saad, M.S. Characterization of beneficial properties of plant growth-promoting rhizobacteria isolated from sweet potato rhizosphere. *Afr. J. Microbiol. Res.* **2009**, *3*, 815–821.

25. Marques, J.M.; da Silva, T.F.; Vollú, R.E.; de Lacerda, J.R.M.; Blank, A.F.; Smalla, K.; Seldin, L. Bacterial endophytes of sweet potato tuberous roots affected by the plant genotype and growth stage. *Appl. Soil Ecol.* **2015**, *96*, 273–281. [CrossRef]

26. Alves, R.P.; Blank, A.F.; Oliveira, A.M.S.; Santana, A.D.D.; Pinto, V.S.; Andrade, T.M. Morpho-agronomic characterization of sweet potato germplasm. *Hort. Bras.* **2017**, *35*, 525–541. [CrossRef]

27. Sakurai, M.; Wasaki, J.; Tomizawa, Y.; Shinano, T.; Osaki, M. Analysis of bacterial communities on alkaline phosphatase genes in soil supplied with organic matter. *Soil Sci. Plant Nut.* **2008**, *54*, 62–71. [CrossRef]

28. Simonet, P.; Grosjean, M.C.; Misra, A.K.; Nazaret, S.; Cournoyer, B.; Normand, P. *Frankia* genus-specific characterization by polymerase chain reaction. *Appl. Environ. Microbiol.* **1991**, *57*, 3278–8326. [PubMed]

29. Poly, F.; Monrozier, L.J.; Bally, R. Improvement in the RFLP procedure for studying the diversity of *nifH* genes in communities of nitrogen fixers in soil. *Res. Microbiol.* **2001**, *152*, 95–103. [CrossRef]

30. Monteiro, J.M.; Vollú, R.E.; Coelho, M.R.R.; Fonseca, A.; Gomes Neto, S.C.; Seldin, L. Bacterial communities within the rhizosphere and roots of vetiver (*Chrysopogon zizanioides* L. Roberty) sampled at different growth stages. *Eur. J. Soil Biol.* **2011**, *47*, 236–242. [CrossRef]

31. Heuer, H.; Krsek, M.; Baker, P.; Smalla, K.; Wellington, E.M.H. Analysis of actinomycete communities by specific amplification of genes encoding 16S rRNA and gel-electrophoretic separation in denaturing gradients. *Appl. Environ. Microbiol.* **1997**, *63*, 3233–3241.

32. Kropf, S.; Heuer, H.; Grüning, M.; Smalla, K. Significance test for comparing microbial community fingerprints using pairwise similarity measures. *J. Microbiol. Meth.* **2004**, *57*, 187–195. [CrossRef]

33. Hammer, Ø.; Harper, D.A.T.; Ryan, P.D. PAST: Paleontological statistics software package for education and data analysis. *Palaeontol. Electron.* **2001**, *4*, 9.

34. Lacerda, J.R.M.; da Silva, T.F.; Vollú, R.E.; Marques, J.M.; Seldin, L. Generally recognized as safe (GRAS) *Lactococcus lactis* strains associated with *Lippia sidoides* Cham. are able to solubilize/mineralize phosphate. *SpringerPlus* **2016**, *5*, 828. [CrossRef]

35. Taketani, R.G.; dos Santos, H.F.; van Elsas, J.D.; Rosado, A.S. Characterisation of the effect of a simulated hydrocarbon spill on diazotrophs in mangrove sediment mesocosm. *Antonie Van Leeuwenhoek* **2009**, *96*, 343–354. [CrossRef] [PubMed]

36. Abdel-Razzak, H.S.; Moussa, A.G.; Abd El-Fattah, M.A.; El-Morabet, G.A. Response of sweet potato to integrated effect of chemical and natural phosphorus fertilizer and their levels in combination with mycorrhizal inoculation. *J. Biol. Sci.* **2013**, *13*, 112–122. [CrossRef]

37. Mukhongo, R.W.; Tumuhairwe, J.B.; Ebanyat, P.; AbdelGadir, A.H.; Thuita, M.; Masso, C. Combined application of biofertilizers and inorganic nutrients improves sweet potato yields. *Front. Plant Sci.* **2017**, *8*, 219. [CrossRef] [PubMed]

38. Oliveira, A.P.; Santos, J.F.; Cavalcante, L.F.; Pereira, W.E.; Santos, M.D.; Oliveira, A.N.; Silva, N.V. Yield of sweet potato fertilized with cattle manure and biofertilizer. *Hort. Bras.* **2010**, *28*, 277–281. [CrossRef]

39. Sharma, S.B.; Sayyed, R.Z.; Trivedi, M.H.; Gobi, T.A. Phosphate solubilizing microbes: Sustainable approach for managing phosphorus deficiency in agricultural soils. *SpringerPlus* **2013**, *2*, 587. [CrossRef]

40. Yoneyama, T.; Terakado-Tonooka, J.; Minamisawa, K. Exploration of bacterial N_2-fixation systems in association with soil-grown sugarcane; sweet potato; and paddy rice: A review and synthesis. *Soil Sci. Plant Nut.* **2017**, *63*, 578–590. [CrossRef]

41. Tarafdar, J.C.; Jungk, A. Phosphatase activity in the rhizosphere and its relation to the depletion of soil organic phosphorus. *Biol. Fert. Soils* **1987**, *3*, 199–204. [CrossRef]

42. Kandeler, E.; Marschner, P.; Tscherko, D.; Gahoonia, T.S. Nielsen NE. Microbial community composition and functional diversity in the rhizosphere of maize. *Plant Soil* **2002**, *238*, 301–312. [CrossRef]

43. Chen, X.; Kou, M.; Tang, Z.; Zhang, A.; Li, H.; Wei, M. Responses of root physiological characteristics and yield of sweet potato to humic acid urea fertilizer. *PLoS ONE* **2017**, *12*, e0189715. [CrossRef]

44. Chung, H.; Park, M.; Madhaiyan, M.; Seshadri, S.; Song, J.; Cho, H.; Sa, T. Isolation and characterization of phosphate solubilizing bacteria from the rhizosphere of crop plants of Korea. *Soil Biol. Biochem.* **2005**, *37*, 1970–1974. [CrossRef]

45. Hill, W.A.; Bacon-Hill, P.; Crossman, S.M.; Stevens, C. Characterization of N_2-fixing bacteria associated with sweet potato roots. *Can. J. Microbiol.* **1983**, *29*, 860–862. [CrossRef]

46. Reiter, B.; Bürgmann, H.; Burg, K.; Sessitsch, A. Endophytic *nifH* gene diversity in African sweet potato. *Can. J. Microbiol.* **2003**, *49*, 549–555. [CrossRef] [PubMed]

47. Terakado-Tonooka, J.; Ohwaki, Y.; Yamakawa, H.; Tanaka, F.; Yoneyama, T.; Fujihara, S. Expressed *nifH* genes of endophytic bacteria detected in field-grown sweet potatoes (*Ipomoea batatas* L.). *Microbes Environ.* **2008**, *23*, 89–93. [CrossRef] [PubMed]

48. Weller, D.M.; Thomashow, L.S. Current challenges in introducing beneficial microorganisms into the rhizosphere. In *Molecular Ecology of Rhizosphere Microorganisms: Biotechnology and Release of GMOs*; O'Gara, F., Dowling, D.N., Boesten, B., Eds.; VCH: New York, NY, USA, 1994; pp. 1–18.

Article

Soil Biological Fertility and Bacterial Community Response to Land Use Intensity: A Case Study in the Mediterranean Area

Mohammad Yaghoubi Khanghahi *, Pasqua Murgese, Sabrina Strafella and Carmine Crecchio

Department of Soil, Plant and Food Sciences, University of Bari Aldo Moro, Via Amendola 165/a, 70125 Bari, Italy; patty.murgese@libero.it (P.M.); sabrina.strafella.bio@gmail.com (S.S.); carmine.crecchio@uniba.it (C.C.)
* Correspondence: mohammad.yaghoubikhanghahi@uniba.it; Tel.: +39-080-5442854

Received: 24 October 2019; Accepted: 8 November 2019; Published: 10 November 2019

Abstract: The current study was performed to investigate the effects of three different long-term land use intensities on adjacent soil plots, namely a winter wheat field, a grass-covered vineyard, and a cherry farm, on soil biochemical, microbial, and molecular parameters. The results showed the maximum content of soil organic matter (SOM) and microbial biomass carbon (MBC) observed in the grass-covered vineyard. Basal respiration (BSR) and the cumulated respiration (CSR) after 25 days of incubation were significantly higher in the grass-covered vineyard and cherry farm, respectively (BSR 11.84 mg CO_2–C kg^{-1} soil d^{-1}, CSR 226.90 mg CO_2–C kg^{-1} soil). Grass-covered vineyard showed the highest soil biological fertility index (BFI) score (20) and ranked in the class IV (good) of soil biological fertility. Cereal field and cherry farm had lower BFI scores and the corresponding BFI class was III (medium). In addition, the maximum ribosomal RNA copy number and the highest abundance of oligotrophic bacterial groups (25.52% Actinobacteria, 3.45% Firmicutes, and 1.38% Acidobacteria) were observed in the grass-covered vineyard. In conclusion, the grass-covered vineyard is a more conservative system and could have a large potential to improve total carbon storage in soil, mainly because of the cover crop residue management and the low soil perturbation through the no-tillage system.

Keywords: bacterial community structure; biological fertility index; land use; microbial biomass; microbial respiration; ribosomal RNA copy numbers

---

1. Introduction

One of the most important negative factors on Mediterranean ecosystems is the conversion of natural landforms into cropland and grassland, due to agricultural development [1]. Nearly 40% of Puglia (a typical Mediterranean region) lands have been converted to farmland for growing olives, table grapes, and almonds over recent centuries [2]. In addition, over 80% of Puglia lands, classified as agricultural land [3], have always been exposed to land degradation because of intensive management practices and land use change [4,5]. Land conversion has become a challenge in Puglia region, where lands are surrounded by soft carbonate rocks and hard calcareous stones [6,7]. However, for a short period of time, soil quality can be improved by the decomposition of rocks because of nutrients released into the soil, but over the long run, land degradation and ecosystem deterioration will occur, which may lead to the farmlands abandonment [2].

Many studies have been focused on examining soil chemical and physical parameters such as soil reaction (pH), soil organic carbon (SOC), cation exchange capacity (CEC), electric conductivity (EC), aggregate stability, bulk density, and soil porosity [8,9]. On the other hand, a proposed minimum dataset for soil quality assessment as quantitative indicators is linked to physical, chemical, and biological parameters or a mix of physical and biological characteristics, such as soil texture, infiltration, and soil bulk density, water holding capacity, SOC, soil microbial biomass, and microbial activity, pH value, total organic carbon, and nitrogen (N), N mineralized under aerobic condition, and extractable N, phosphorus (P), and potassium (K) [10,11].

Therefore, a comprehensive indicator system is required to be used for soil monitoring among all types of biological indicators and biogeochemical cycles [12]. In this regard, the soil Biological Fertility Index (BFI) was introduced by Renzi et al. [13] according to some biochemical parameters of soil, including SOM, microbial biomass carbon (MBC), basal respiration at the last day of incubation (BSR), cumulated respiration during the incubation period (CSR), metabolic quotient (qCO$_2$), and mineralization quotient (qM). It has already been reported that the BFI indicator may be more efficient than microbial biomass and activity alone and can be used to assess soil quality at varying levels of human disturbance [12,13].

There have been relatively limited researches dealing with the impacts of different types of land use intensities on microbial and biochemical soil properties in the Mediterranean-type ecosystems and extremely rarely addressed by ecological researchers in term of the application of precise indicators in complex Mediterranean mosaic landscapes [12,14]. Therefore, the present research aimed to investigate the effects of different land use intensities, which are frequently adopted in the agro-ecosystems of the Mediterranean area, on soil chemical, biochemical and microbial parameters. Three different long-term land use intensities on adjacent soil plots, recently converted from natural landform to farmland by rock fragmentation and intensive agricultural management, were investigated. Namely, soil plots were a winter wheat field (soil tillage, residues removal, integrated pest management, no irrigation, chemical fertilization), a grass-covered vineyard (no tillage, residues left on the field, integrated pest management, drip irrigation with continuous mixed chemical and organic fertilization), and a cherry farm (soil tillage, residues removal, drip irrigation with continuous chemical fertilization). Our hypothesis is that the long-term land use changes soil biological fertility, evaluated by the Biological Fertility Index (BFI) and composition and abundance of soil bacterial community.

2. Materials and Methods

The study was performed in a Mediterranean soil ecosystem in Puglia region, Italy (Figure 1) (Turi, located at 40°91′ N, 17°04′ E, altitude of 247 m above the average of sea level, with long-term annual precipitation range from approximately 550 mm to 600 mm and annual temperature range from 4 to 30 °C) with Mediterranean climate conditions according to the Domarten classification.

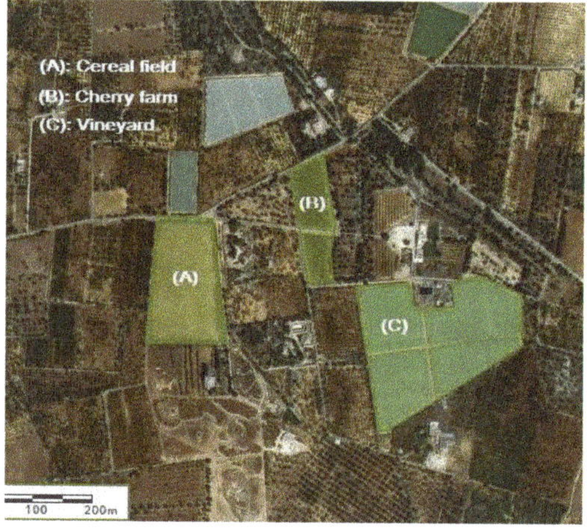

Figure 1. The area located at Turi in Puglia region, Southern Italy.

Three land-use intensities, namely cereal field (winter wheat), grass-covered vineyard, and cherry farm plots, were compared in adjacent plots. The land use descriptions and management are presented in Table 1.

Table 1. Land use management.

Land Use	Area (Ha)	Cultivation History (Year)	Management	Irrigation and Fertilization
Cereal field	2.4	7	Soil tillage (40 cm), residues removal from the field, integrated pest management,	No irrigation (dryland farming), chemical fertilization before seeding
Grass-covered vineyard	4.9	10	No tillage, residues left on the field, integrated pest management,	Drip irrigation with continuous mixed fertilization (chemical–organic)
Cherry farm	1.2	10	Soil tillage (40 cm), residues removal from the farm, integrated pest management	Drip irrigation with continuous chemical fertilization

The area was a natural landform with carbonates, clay, and iron oxides accumulation in depth (Chromic and Calcic Luvisols) that has been converted to farmland by rock fragmentation and intensive agricultural practices about 10 years ago and cultivated continuously. The geological characteristics of the plots (Luvisols from limestone) are similar and typical for this region.

Three composite soil samples from each land use were taken randomly from the upper soil layer in June 2018. A minimum of five sub-samples were collected, mixed together on site and considered as a composite soil sample. Samplings were carried out at depth of 0–20 cm, since the maximum microbial biomass is devoted to the upper soil horizons [15] and microbial activity is sensitive to changed land uses [16]. The fraction passing a 2 mm sieve was transported refrigerated at 4 °C.

2.1. Measurements

Soil pH was measured in an aqueous matrix in 1:2.5 soil: water suspension [17] using pH meter (XS Instruments pH 50, Carpi, Italy), soil texture according to the USDA classification [18] by the relative amounts of sand, silt, and clay in the fine earth fraction includes all soil particles that are less than 2 mm. The Walkley and Black [19] method was performed to measure total soil organic carbon (SOC) and the van Bemmelen factor (=1.724) was used to calculate the amount of soil organic matter (SOM) with the following equation [20]:

$$SOM = SOC \times 1.724 \tag{1}$$

2.2. Soil Microbial Biomass

The chloroform fumigation extraction method [21] was used to estimate the microbial biomass carbon (MBC). In details, the air-dried soil was pre-incubated for five days at 30 °C in open glass jars at field capacity. Three replicates of each soil (12.5 g) were fumigated with free ethanol chloroform for 24 h in vacuum desiccators at dark condition, and other three replicates were not fumigated as blank. Both the fumigated and non-fumigated soil samples were extracted with 0.5 M K_2SO_4 for 30 min and the resulting extracts were filtered. Potassium dichromate method [21] was used to measure MBC in filtrates.

2.3. Microbial Respiration

Soil microbial respiration was measured according to Isermeyer [22]. Twenty-five-gram soil samples were incubated in closed glass jars under dark conditions at field capacity and 30 °C. The CO_2 evolved was trapped by 0.2 N NaOH and measured by titration of the excess NaOH with 0.2 N HCl. The jars without soil maintained at the same way served as a CO_2 blank. The difference of consumed volume of HCl between the samples and the blank in titration was considered to estimate the amount of CO_2 evolution by soil microorganisms. Cumulated microbial respiration (CSR) is the respiration rate during 25 days of incubation period in mg CO_2–C kg^{-1} soil. Basal respiration rate (BSR) is the respiration at the last day of incubation (The 25th day) expressed in mg CO_2–C kg^{-1} soil d^{-1}.

The microbial metabolic quotient (qCO_2) indicates the amount of CO_2–C produced per unit MBC and was calculated from BSR per unit MBC in mg CO_2–C 10^{-2} h^{-1} mg MBC^{-1} [23]. The mineralization quotient (qM) is expressed in% and calculated as the ratio of the CSR to SOC. The qM represents the efficiency of microflora to metabolize the soil organic carbon [24]. The efficiency of microbes to decompose organic carbon (CUE) was estimated as the ratio of microbial biomass carbon to soil organic carbon [25].

2.4. Biological Fertility Index

The Biological Fertility Index (BFI) is a comprehensive indicator by composing six biological variables, such as SOM, BSR, CSR, MBC, qCO_2, and qM. Five intervals of values were set for each parameter, and scores increasing from 1 to 5 were assigned to each interval (Table 2) based on the available evidences from previous research [13]. The algebraic sum of scores for all the parameters providing the proposed classes of biological fertility are shown in Table 2

Table 2. Scores of the intervals of values for the different parameters.

Parameters	Scores				
	1	2	3	4	5
Soil organic matter, SOM (%)	<1.0	$\geq$1.0 $\leq$1.5	>1.5 $\leq$2.0	>2.0 $\leq$3.0	>3.0
Basal soil respiration, BSR (mg CO_2–C kg^{-1} soil d^{-1})	<5	$\geq$5 $\leq$10	>10 $\leq$15	>15 $\leq$20	>20
Cumulative soil respiration, CSR (mg CO_2–C kg^{-1} soil)	<100	$\geq$100 $\leq$250	>250 $\leq$400	>400 $\leq$600	>600
Microbial biomass carbon, MBC (mg C kg^{-1} soil)	<100	$\geq$100 $\leq$200	>200 $\leq$300	>300 $\leq$400	>400
Metabolic quotient, qCO_2 (mg CO_2–C 10^{-2} h^{-1} mg MBC^{-1})	$\geq$0.4	$\geq$0.4 $\leq$0.3	<0.3 $\geq$0.2	<0.2 $\geq$0.1	<0.1
Mineralization quotient, qM (%)	<1.0	$\geq$1 $\leq$2	>2 $\leq$3	>3 $\leq$4	>4
Classes of the Biological Fertility Index (BFI)					
Fertility class	I	II	III	IV	V
	Stress	Pre-stress (alarm)	Medium	Good	High
BFI scores sum	6	7–12	13–18	19–24	25–30

2.5. Soil DNA Extraction

The FastDNA® SPIN Kit for Soil (MP Biomedicals, Solon, CA, USA) was used for DNA extraction. Briefly, 0.5 g of soil was added into Lysing Matrix E tubes, and the samples were homogenized in a FastPrep® at setting 6.0 m s^{-1} for 40 s. All extraction steps were performed according to the manufacturer's instructions for kit. Nucleic acid concentration was evaluated using a Nanodrop spectrophotometer (ND-1000, EuroClone, Milan, Italy).

2.6. Real-Time Quantitative PCR Analysis (qPCR)

qPCR was performed in triplicate to evaluate bacterial 16S rRNA gene copy numbers using an Applied Biosystems 7500 detection system (Applied Biosystems, Foster City, CA, USA) with 341 primer pairs, forward (5'-CCTACGGGNGGCWGCAG-3') and 785 primer pairs, reverse (5'-GACTACHVGGGTATCTA ATCC-3') [26]. The reaction mixture (20 μL) and amplification conditions were performed based on the methods of Pascazio et al. [27]. Reference strain DNA of *Azospirillum irakense* [28] was used to create the standard curve and the gene copy number in the samples was calculated using the regression equation that related the cycle threshold (Ct) value to the number of copies in the standard curve [29].

2.7. Analyses of 16S rRNA Gene Sequence, Species Richness and Diversity

Aliquots of the extracted DNA were sent to the IGA Technology Service in Udine (Italy) for metagenomic analyses; the sequencing was performed through the MiSeq Illumina system platform. QIIME software (version 1.9.1) was used to perform bacterial community analysis as fully described by Kuczynski et al. [30]. There were two amplification steps in the library workflow: an initial PCR amplification using locus specific PCR primers and a subsequent amplification that integrates relevant flow-cell binding domains and unique indices (NexteraXT Index Kit, FC-131-1001/FC-131-1002). This method was used to amplify the variable V3 and V4 regions of the 16S rRNA gene aiming to characterize bacterial community compositions. Operational taxonomic units (OTUs) were identified using a cut-off of 97% similarity. Singletons, non-bacterial OTUs were removed, and the OTU abundance levels were normalized based on the sample with the least number of sequences.

To estimate the bacterial α-diversity, CHAO1, Shannon and Simpson indices and good-coverage were calculated using the QIIME software based on 10,000 sequences per sample. Rarefaction curves endpoints and normalization of counts for diversity analysis was set to 50% of the target sequencing coverage (i.e., for 100,000 fragments a cutoff of 50,000 fragments were applied). Bacterial taxonomic assignment was conducted using the Ribosomal Database Project (RDP) Naïve Bayesian classifier and reference database with a minimum confidence threshold of 50%.

2.8. Statistical Analysis

All statistical analyses including the least significant difference (LSD) test, performed to compare the differences among parameters means at a 5% level, correlation (Pearson, two-tailed), regression (multiple linear regression, backward method) and cluster (Ward's method) analysis were done using SPSS software (Statistical Product and Service Solutions, version 16, IBM, New York, NY, USA).

3. Results

3.1. Chemical and Microbial Parameters

Results of the soil chemical parameters and microbial activity in different land use are summarized in Table 3. There was no difference among the three land uses in terms of soil pH and texture. However, cherry farm soil had a slightly lower sand and silt content and higher pH than the other soils. The results showed that soil organic carbon (SOC) of the samples from grass-covered vineyard and from cereal field was significantly greater than that from the cherry farm. In particular, SOC was 1.82% and 1.69% in the grass-covered vineyard and cereal field, respectively, which was 1.33 and 1.24 times more than that in the cherry field (Table 3). Accordingly, the maximum soil organic matter (SOM) was recorded in grass-covered vineyard, while the lowest was found in cherry farm (3.13% and 2.35%, respectively). The maximum values of microbial biomass (MBC) was observed in the grass-covered vineyard soil (201.92 μg C g^{-1} soil), which was about 2.45% and 8.29% more than that in cereal and cherry farm soils, respectively. Basal soil respiration (BSR) and the cumulated respiration (CSR) after 25 days of incubation were significantly higher in grass-covered vineyard and cherry farm, respectively (BSR 11.84 mg CO$_2$–C kg^{-1} soil d^{-1}, CSR 226.90 mg CO$_2$–C kg^{-1} soil) than in the cereal field (BSR 9.73 mg CO$_2$–C kg^{-1} soil d^{-1}, CSR 153.91 mg CO$_2$–C kg^{-1} soil). The highest and significantly different value of efficiency of microbes to decompose organic matter (CUE) was found in cherry farm (138.08 μg C g^{-1} soil). The highest amount of metabolic quotient (qCO$_2$) were detected in grass-covered vineyard (0.058 mg CO$_2$–C 10^{-2} h^{-1} mg MBC^{-1}) and cherry farm (0.056 mg CO$_2$–C 10^{-2} h^{-1} mg MBC^{-1}), which were about 16% and 12% higher than that in the cereal field. We also found statistically significant differences among land uses in term of carbon mineralization quotient (qM). These values varied from 0.91% in the cereal field to 1.66% in the cherry farm.

Table 3. Chemical and microbial parameters under different land use.

Parameter	Unit	Land Use		
		Cereal Field	Grass-Covered Vineyard	Cherry Farm
Soil pH		7.95 ± 0.04	7.95 ± 0.01	7.98 ± 0.04
Soil texture			Sandy Clay Loam	
SOC	%	1.69 ± 0.17 a	1.82 ± 0.13 a	1.36 ± 0.11 b
SOM	%	2.91 ± 0.29 a	3.13 ± 0.22 a	2.35 ± 0.19 b
MBC	$\mu g\ C\ g^{-1}$ soil	197.08 ± 18.76 a	201.92 ± 20.81 a	186.45 ± 14.19 b
CUE	$\mu g\ C\ g^{-1}$ soil	117.06 ± 1.14 b	110.53 b ± 4.96	138.08 a ± 12.98
BSR	mg CO_2–C kg^{-1} soil d^{-1}	9.73 ± 0.87 b	11.84 ± 0.66 a	10.51 ± 1.11 ab
CSR	mg CO_2–C kg^{-1} soil	153.91 ± 10.77 b	214.49 ± 13.69 a	226.90 ± 18.88 a
qCO_2	mg CO_2–C $10^{-2}\ h^{-1}$ mg MBC^{-1}	0.050 ± 0.004 b	0.058 ± 0.005 a	0.056 a ± 0.004
qM	%	0.91 ± 0.08 c	1.17 ± 0.10 b	1.66 ± 0.15 a
Nucleic acid concentration	$ng\mu L^{-1}\ 0.5\ g^{-1}$	154.8 ± 6.01 b	160.6 ± 5.49 a	143.7 ± 5.94 c
16S rRNA- CN	g^{-1} DS	$8.06 \pm 1.28 \times 10^7$ a	$1.82 \pm 0.51 \times 10^7$ c	$5.18 \pm 0.32 \times 10^7$ b

Means (± standard error) of each parameter followed by a similar letter are not significantly different based on the least significant difference (LSD) test at the 5% probability level. SOC: total organic carbon, SOM: total organic matter, MBC: microbial biomass C, CUE: microbial carbon use efficiency, BSR: basal soil respiration, CSR: cumulative soil respiration, qCO_2: metabolic quotient, qM: mineralization quotient, 16S rRNA- CN: Bacterial 16S rRNA gene copy numbers.

Regression of cubic equation models significantly fitted ($P < 0.01$) correlations between microbial respiration rate and time period of incubation. The coefficients of determination (R^2) of the equations were 0.98, 0.95, and 0.99 in the cereal field, grass-covered vineyard, and cherry farm, respectively (Table 4, Figure 2).

Table 4. Regression coefficient for determining the relationship between the microbial respiration rates in different land use after an incubation period of 25 days.

Land Use	Equation	R Square	Constant	b1	b2	b3
Cereal field	Cubic	0.98 **	18.28	1.57	−0.18	0.004
Grass-covered vineyard	Cubic	0.95 **	38.79	−0.75	−0.12	0.004
Cherry farm	Cubic	0.99 **	61.92	−8.11	0.44	−0.008

** corresponds to significance at $p < 0.01$.

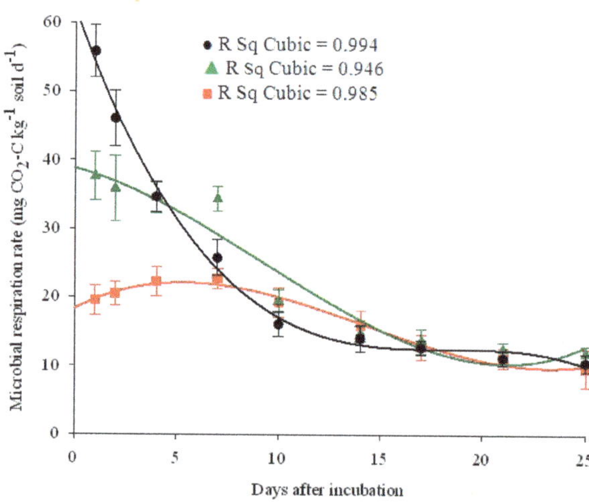

Figure 2. Respiration rate during 25 days of incubation in cereal field (■), grass-covered vineyard (▲), and cherry farm (●).

3.2. Soil Biological Fertility Index

The soil Biological Fertility Index (BFI) was calculated to determine the overall effect of land-use systems on soil quality. The BFI results and the average scores of the parameters with respect to land use are given in Table 5. Grass-covered vineyard showed the highest BFI score (20) and ranked in class IV (good) of soil biological fertility, which is assumed to be a sustainable ecosystem for the long-term. The other two managements (cereal field and cherry farm) had lower BFI scores (16 and 18, respectively), corresponding to a BFI of class III (medium) (Table 5).

Table 5. Scores of the soil parameters and biological fertility index for the different land uses.

Land Use	SOM	BSR	CSR	MBC	qCO_2	qM	BFI Score	BFI Class
Cereal field	4	2	2	2	5	1	16	III (medium)
Grass-covered vineyard	5	3	2	3	5	2	20	IV (good)
Cherry farm	4	3	2	2	5	2	18	III (medium)

SOM: total organic matter, MBC: microbial biomass C, BSR: basal soil respiration, CSR: cumulative soil respiration, qCO_2: metabolic quotient, qM: mineralization quotient, BFI: biological fertility index.

The multiple linear regressions were used to explain which of the biological parameters could be more useful to predict the BFI under different land uses. In order to achieve this purpose, backward elimination method was used, and the results are shown in Table 6. Accordingly, SOM and CSR were identified as two independent variables which had most impact on BFI (dependent variable). The rankings of the parameters were SOM > CSR > qM > qCO_2 > BSR > MBC. This multiple linear regression model with two explanatory variables (SOM and CSR) had an R^2 of 0.916. Therefore, 91.6% of the variation in%BFI can be explained by this model and variables (Table 6).

Table 6. Multiple linear regressions (backward method) among BFI (dependent variable) and six biological variables (independent variables) under three land uses.

Model	Independent Variables	Standardized Coefficients (Beta)	t	Significance Level
	(Constant)		1.059	0.401
	SOM	1.119	0.932	0.450
	BSR	0.501	0.603	0.608
1	CSR	1.230	1.352	0.309
	MBC	−1.411	−0.550	0.638
	qCO_2	−1.448	−0.625	0.596
	qM	−0.466	−0.980	0.430
	(Constant)		3.756	0.033
	SOM	0.519	1.181	0.323
	BSR	0.208	0.372	0.735
2	CSR	0.792	2.061	0.131
	qCO_2	−0.206	−0.454	0.680
	qM	−0.426	−1.035	0.377
	(Constant)		4.827	0.008
	SOM	0.650	2.782	0.050
3	CSR	0.896	3.835	0.019
	qCO_2	−0.046	−0.363	0.735
	qM	−0.445	−1.229	0.286
	(Constant)		5.318	0.003
	SOM	0.641	3.035	0.029
4	CSR	0.903	4.269	0.008
	qM	−0.484	−1.544	0.183
	(Constant)		5.703	0.001
5	SOM	0.931	8.765	0.000
	CSR	0.612	5.761	0.001

Model Summary				
Model	R	R^2	Adjusted R^2	Std. Error of the Estimate
1	0.983	0.966	0.863	0.46969
2	0.980	0.961	0.895	0.41147
3	0.979	0.959	0.918	0.36447
4	0.978	0.957	0.932	0.33132
5	0.968	0.937	0.916	0.36752

SOM: total organic matter, MBC: microbial biomass C, BSR: basal soil respiration, CSR: cumulative soil respiration, qCO_2: metabolic quotient, qM: mineralization quotient.

3.3. Nucleic Acid Concentration

The amount of nucleic acid was affected by different land use intensities. The highest amount of nucleic acid was extracted from grass-covered vineyard soil samples (160.6 $ng\mu L^{-1}$ 0.5 g^{-1}) and the lowest value from cherry farm soil samples (143.7 $ng\mu L^{-1}$ 0.5 g^{-1}) (Table 3).

3.4. Bacterial Quantification

Bacterial quantification was performed on DNA extracts using a quantitative real-time polymerase chain reaction (qPCR) analysis. The bacterial ribosomal RNA gene copy numbers ranged from 1.82 to 8.06×10^7 g^{-1} soil. Land use intensities had a significant effect on the ribosomal gene abundances, which was higher in the cereal field than that in the DNA extracted from grass-covered vineyard and cherry farm soils (Table 3).

3.5. Correlation Coefficients among Parameters

The Pearson correlation coefficients showed that there were significant positive correlations between MBC and SOM, BSR and qCO_2, CSR and qM ($p < 0.05$), and negative correlations between MBC and qCO_2 ($p < 0.05$) and qM and SOM ($p < 0.05$) is soil under different managements. In addition, there was a significant positive correlation between the nucleic acid concentration and SOM and MBC in soil samples under different land uses. We observed a negative correlation ($p < 0.05$) between nucleic acid concentration and qM. A significantly ($p < 0.05$) negative correlation was also observed between the rRNA gene copy numbers and soil microbial respiration (both basal and cumulative respirations) (Table 7).

Table 7. Pearson correlation coefficients (r) between biological and molecular variables under different land uses ($n = 9$).

Parameters	SOM	BSR	CSR	MBC	qCO_2	qM	Nucleic Acid Concentration	16S rRNA- CN
SOM	1							
BSR	0.25 ns	1						
CSR	−0.27 ns	0.60 ns	1					
MBC	0.78 *	−0.10 ns	−0.09 ns	1				
qCO_2	−0.41 ns	0.68 *	0.44 ns	−0.79 *	1			
qM	−0.76 *	0.27 ns	0.76 *	−0.58 ns	0.59 ns	1		
Nucleic acid concentration	0.91 **	0.31 ns	−0.21 ns	0.68 *	−0.33 ns	−0.69 *	1	
16S rRNA- CN	−0.15 ns	−0.71 *	−0.68 *	−0.00 ns	−0.45 ns	−0.25 ns	−0.27 ns	1

* and **: Significant at $p < 0.05$ and $p < 0.01$ levels, respectively. ns: not significant. SOM: total organic matter, MBC: microbial biomass C, BSR: basal respiration, CSR: cumulative soil respiration, qCO_2: metabolic quotient, qM: mineralization quotient, 16S rRNA CN: Bacterial 16S rRNA gene copy numbers.

3.6. Bacterial α Diversity and Community Composition

A total of 479,160 high-quality sequences were detected with a 300-bp read length. The minimum Good's coverage value was 0.99 according to the similarity cut-off of 97%, meaning that a sufficient number of reads were obtained to evaluate bacterial diversity. There were no statistically significant differences among land uses in term of bacterial α-diversity (Table 8), although the grass-covered vineyard non-significantly increased CHAO1 and Shannon indices, as compared to the other ecosystems.

Table 8. Estimation of α-diversity indexes for bacterial communities under different land uses (sequence count: 10,000).

Land Use	CHAO1	Simpson	Shannon	Goods-Coverage
Cherry farm	4882.3 ± 108.95 a	0.85 ± 0.00 a	9.93 ± 0.8 a	0.99 ± 0.00 a
Cereal field	4558.8 ± 97.04 a	0.87 ± 0.00 a	9.84 ± 0.05 a	0.99 ± 0.00 a
Grass-covered vineyard	4973.2 ± 61.94 a	0.85 ± 0.00 a	10.14 ± 0.4 a	0.99 ± 0.00 a

Means (± standard error) in each column followed by similar letter are not significantly different based on the least significant difference (LSD) test at 5% probability level.

Relative abundances (%) of bacteria at the phylum (> 1%) and family levels (> 2%) are presented in Table 9. Proteobacteria was the dominant phylum in all soil samples of land uses, which varied from 26.38% and 26.43% (in the cherry farm and grass-covered vineyard, respectively) to 29.29% (in the cereal field) of the total sequences, followed by Actinobacteria (22.88%–25.51%) and Bacteroidetes (6.88%–8.82%). The maximum abundance of Actinobacteria, Firmicutes and Acidobacteria was detected in the samples of the grass-covered vineyard. Moreover, the abundance of Planctomycetes and Gemmatimonadetes were also higher in the cherry farm than those in other ecosystems.

Table 9. Relative abundance of (**A**) bacterial phyla (relative abundance > 1%) and (**B**) families (relative abundance > 2%) under different land uses.

(A)	Land Use		
Phylum	**Cherry Farm (%)**	**Cereal Field (%)**	**Grass-Covered Vineyard (%)**
Proteobacteria	26.38 ± 0.58 b	29.29 ± 0.79 a	26.43 ± 0.23 b
Actinobacteria	22.88 ± 0.25 b	23.71 ± 1.05 b	25.51 ± 1.41 a
Bacteroidetes	7.58 ± 0.99 a	8.82 ± 0.46 a	6.88 ± 0.35 a
Planctomycetes	5.03 ± 0.19 a	4.13 ± 0.21 ab	3.98 ± 0.28 b
Verrucomicrobia	4.67 ± 0.11 ab	5.28 ± 0.51 a	3.09 ± 0.11 b
Chloroflexi	4.33 ± 0.66 a	3.85 ± 0.27 a	3.89 ± 0.23 a
Firmicutes	3.25 ± 0.39 ab	2.32 ± 0.20 b	3.45 ± 0.14 a
Gemmatimonadetes	2.00 ± 0.04 a	1.31 ± 0.05 c	1.64 ± 0.10 b
Acidobacteria	1.23 ± 0.07 b	0.95 ± 0.03 c	1.38 ± 0.02 a

(B)				
Phylum	**Family**	**Cherry Farm (%)**	**Cereal Field (%)**	**Grass-Covered Vineyard (%)**
Actinobacteria	*Rubrobacteraceae*	7.20 ± 0.35 b	6.09 ± 0.27 b	8.68 ± 0.28 a
Proteobacteria	*Bradyrhizobiaceae*	2.33 ± 0.29 a	2.56 ± 0.08 a	2.78 ± 0.13 a
Firmicutes	*Bacillaceae*	2.10 ± 0.31 a	1.23 ± 0.14 b	2.02 ± 0.16 a
Actinobacteria	*Solirubrobacteraceae*	2.56 ± 0.58 a	1.87 ± 0.02 b	2.42 ± 0.19 a
Bacteroidetes	*Chitinophagaceae*	2.75 ± 0.24 b	2.81 ± 0.02 b	3.87 ± 0.32 a
Planctomycetes	*Gemmataceae*	2.55 ± 0.04 a	1.26 ± 0.10 b	1.61 ± 0.13 b
Gemmatimonadetes	*Gemmatimonadaceae*	2.00 ± 0.04 a	1.31 ± 0.05 c	1.64 ± 0.10 b
Proteobacteria	*Rhodospirillaceae*	1.66 ± 0.08 b	1.97 ± 0.04 ab	2.59 ± 0.17 a
Proteobacteria	*Sinobacteraceae*	2.04 ± 0.08 a	2.11 ± 0.22 a	1.37 ± 0.12 b
Verrucomicrobia	*Pedosphaeraceae*	2.31 ± 0.33 a	1.65 ± 0.05 b	8.68 ± 0.28 a

Means (± standard error) in each phylum or family followed by similar letter are not significantly different based on the least significant difference (LSD) test at the 5% probability level.

A total of 10 abundant families (relative abundance > 2%) were identified (Table 9). Rubrobacteraceae represented a range of percentage from 6.09% (cereal field) to 8.68% (grass-covered vineyard) of the total sequences, followed by Bradyrhizobiaceae (2.33–2.78%) and Bacillaceae (1.23–2.10%). Bacillaceae and Solirubrobacteraceae were 2.10% and 2.56% in the cherry farm, and 2.02% and 2.42% in the grass-covered vineyard, respectively, which were significantly higher than those percentages in the cereal field (1.23 and 1.78%, respectively). The relative abundance of Gemmataceae, Gemmatimonadaceae, Sinobacteraceae, and Pedosphaeraceae were also significantly influenced by the cherry farm system. In opposite, the highest abundance of Chitinophagaceae and Rhodospirillaceae were obtained from the samples of the grass-covered vineyard, which was significantly higher than the corresponding abundances in other fields.

At the genus level, the majority of bacteria among all fields belonged to the genera *Rubrobacter*, *Bacillus*, *Gemmatimonas*, *Gemmata*, *Steroidobacter*, and *Pedosphaera* (Figure 3). Cluster analysis showed that the bacterial community composition of the cherry farm and grass-covered vineyard clustered into one group, with a distance of 2.53%, which was separated from the cereal field (Figure 3).

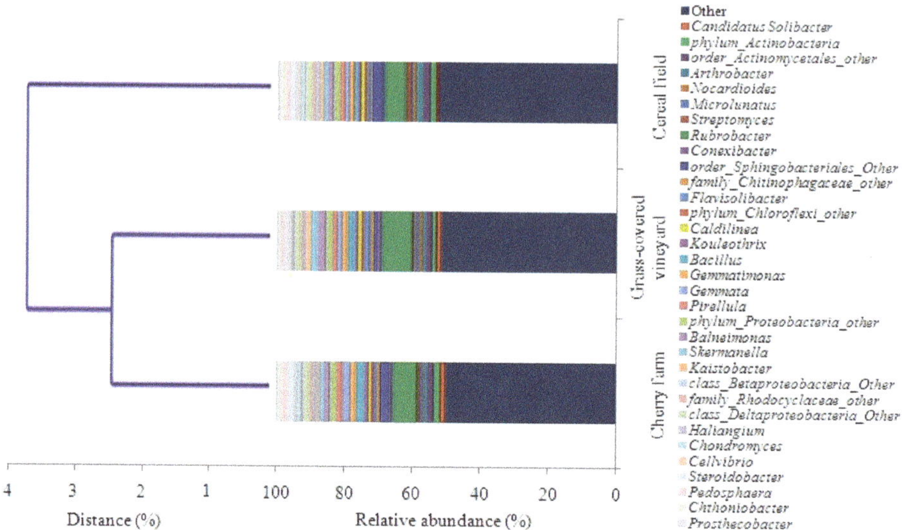

Figure 3. Analysis (Ward's method) of the 16S rRNA composition of soil bacterial communities at the genera level.

4. Discussion

Our results showed that the SOM, MBC, and BSR were highest in the topsoil of the grass-covered vineyard. The increase in the SOM amount in these soil samples was likely due to a mix of different reasons, some of them already demonstrated by other authors, namely, the type of cropping systems and tillage intensity, e.g., the reduction of soil disturbance through no-tillage [31], the management of crop residues [32], and of cover crops [33], both practices increasing the plant inputs returned to the soil. In fact, crops' shoot and root biomass of cover crops and grape trees were left on the field, in contrast with the cereal field and cherry farm, where residue removal and crop harvesting decreased organic inputs. These findings are in agreement with Ramesh et al. [34], who found a considerable impact of land use changes on SOC and reported that chemical and biochemical soil parameters were constant under low-intensity farming practices. A similar finding was also reported by Safaei et al. [35] who showed that the SOM and soil quality in natural ecosystems were considerably affected by land uses. Increasing MBC is the consequence of high-concentration C substrates available to the soil microbial communities [36] and of a significant positive trend in the quantity of organic C in the natural or low-disturbance ecosystems over the long run [12], such as the grass-covered vineyard in the present research. Similarly, an increase in MBC has already been found with decreased levels of human activities i.e., no-tillage practices [37] and straw returning [38] by changing the soil physicochemical environment, such as soil water content, porosity, and bulk density.

According to the results, the rankings were grass-covered vineyard > cherry field > cereal field for BSR, and cherry farm > grass-covered vineyard > cereal field for CSR. This means that the microbial respiration was increased by higher organic C inputs to the soil from crop residue management in the grass-covered vineyard. On the other hand, soil respiration and microbial activity are strongly influenced by soil water content [39]. Therefore, in the present study, lower rates of basal and cumulative respiration in the cereal field as compared to others may be explained by dry land farming and lower water content.

The ranking for CUE was cherry field > cereal field > grass-covered vineyard. CUE is a principal parameter used to estimate soil C dynamics and to understand the destiny of C resources by partitioning organic C between MBC and mineralized CO_2 [25]. In the cherry farm, higher CUE indicated an increased decomposition, resulting in reduction of plant and microbial organic matter in soil [40].

The highest amounts of metabolic quotient (qCO_2) were detected in grass-covered vineyard (0.058 mg CO_2–C 10^{-2} h^{-1} mg MBC^{-1}) and cherry farm (0.056 mg CO_2–C 10^{-2} h^{-1} mg MBC^{-1}), which were about 16 and 12% higher than that in cereal field. It has been reported that stress conditions may cause an increment in the qCO_2 which is related to a microbial stress, while low qCO_2 may indicate more desirable conditions for microbial survival and more carbon available for biomass production [41]. High qCO_2 values could also be related to changes in the bacterial-to-fungal ratio [42]. In this regard, Nsabimana et al. [43] determined qCO_2 and CUE parameters to illustrate how the soil microbial activity and community composition were affected by land uses. In addition, we also found statistically significant differences among land uses in term of carbon mineralization quotient (qM). These values varied from 0.91% in the cereal field to 1.66% in the cherry farm, meaning that the cherry farm had better microbial C metabolism efficiency [44]. Lower amounts of qM in the cereal field than that in other ecosystems are related to lower ratios of easily mineralizable organic matter to stable organic matter in the soil [24].

There were statistically significant relationships between microbial respiration rate and the time period of incubation as cubic equation models. These results are also in agreement with Birge [45] who reported that the microbial respiration rate declined over the course of the incubation due to lack of organic C resource. Dungait et al. [46] suggested that the availability of soil organic C to microbial decomposers is one of the main factors that can influence and limit microbial respiration. This hypothesis has been supported by Birge [45] who stated that the SOM availability, and not the microbial biomass content, could limit the respiration rate.

Our results showed that the highest value of BFI was obtained from grass-covered vineyard. It has been previously reported that soil fertility is proportional to the soil microbial activity and SOM, as well as altering environmental conditions and land uses [47,48]. It has also been proved that crop residues and cover crops increase soil microbial biomass and activity and improve SOC sequestration in the long-term [49]. Therefore, cover crop residues stimulated cumulative carbon mineralization and improved SOC cycling in vineyard ecosystem, resulting in high level of BFI. In fact, grass-covered vineyard, being a more conservative system, has a large potential to improve total carbon storage in soil, given the organic input and the lower soil disturbance from tillage operations. On the other hand, BFI score in cereal field was lower than that in other land uses. This was very likely due to the frequent tillage practices and the removal of crop residues from the field, as well as to the dryland farming that can lead to a low soil biological fertility. These findings highlight the need for proper sustainable land management to prevent soil degradation [50], since the sustainability of the dryland ecosystem and its agricultural production depends strongly on proper and effective land-use and management [51]. Thus, it is quite clear that the conventional tillage with the crop-fallow system should be avoided in dryland cropping systems because of consequently reduced soil fertility and degraded chemical properties [52].

According to the evaluation by multiple linear regressions, SOM and CSR seem to be more useful to predict the BFI under different land uses. Similar results were presented by Renzi et al. [13], who found that the BFI had a significant relationship with SOM, CSR, MBC, and BSR, and increased linearly with all composing variables except for qM and qCO_2, and particularly that MBC and CSR were the most important parameters contributing to the BFI.

The amounts of extracted nucleic acid were affected by different land use intensities, too. A similar finding was reported by Agnieszka et al. [53], who demonstrated that the soil DNA content was significantly influenced by land uses. It is well recognized that the soil DNA concentration is higher in soils with higher soil organic carbon [40,54]; in this regard, we found a significant positive correlation between changes in the nucleic acid concentration and SOM and MBC in soils under different land uses. Other authors reported a strong correlation between microbial biomass carbon and extracted DNA in soil [53], this relationship being affected by land use intensity and clearly indicating that microbial ecophysiology can be directly linked to soil carbon storage potential [40].

The composition of the soil bacterial communities in the cereal ecosystem, a management characterized by a higher disturbance and lower values of biochemical and microbiological parameters, shifted to species with a higher number of 16S rRNA gene copies. It has been reported that the rRNA gene copy numbers are linked to the bacterial life strategy [55], basically classified into two main categories, copiotrophic and oligotrophic, according to their response to resource availability and life strategies [56]. Therefore, higher ribosomal RNA copy numbers detected in the cereal field than that in other land uses, as indicated by the qPCR assay, indicate an increase in fast-growing r-strategists' taxa (copiotrophic species), due to decreased microbial respiration [55,57].

Bioinformatics analyses of the 16S sequencing data completely matched with the rRNA gene copy numbers. Accordingly, the abundance of oligotrophic bacterial groups (25.52% Actinobacteria, 3.45% Firmicutes, and 1.38%Acidobacteria) was higher in the grass-covered vineyard, while the rRNA gene copy numbers were lower than for other land uses. Oppositely, the cereal field had the highest abundance of copiotrophic bacteria (29.29% Proteobacteria and 8.82% Bacteroidetes) which was 10.8 and 28.2% more than grass-covered vineyard and 11.0 and 11.6% higher than cherry farm, respectively. The adaptive capacities of these species permit a successful competition with other bacterial groups, particularly when the levels of soil disturbance or tillage intensity are higher. In the cereal field, we hypothesize that soil disturbance repressed microbial respiration by favoring r-strategists' taxa with faster growth rates and limited capabilities to degrade recalcitrant organic matter, as indicated by a significant negative correlation between rRNA copy number and soil microbial basal as well as cumulative respiration.

CHAO1, Shannon and Simpson diversity indices indicated that the whole bacterial communities were very similar, no matter the land use intensities. This finding has already been proved by many authors [58–60], who reported that there was no effect of land use and agricultural practices on the diversity indices tested (e.g., CHAO1 and Shannon). Our research supports the hypothesis that different land uses do not change species richness and heterogeneity, although significant differences in some bacterial groups were found, very likely because of the high microbial resilience and/or resistance [61].

The results of Pearson correlation coefficients agree with the results of Renzi et al. [13], who found a significant relationship between SOM and MBC, BSR and qCO_2, and CSR and qM ($p < 0.001$) based on Pearson correlation analyses. Similarly, Li et al. [62] and Malik et al. [40] reported that the SOM significantly correlated with MBC in different land use. It has also been reported that soil microbial communities were affected by quality and quantity of soil organic carbon [38]. Van Wesemael et al. [41] reported that the high qCO_2 values represent the high energy required to keep up the microbial biomass and indicate a stressful condition for microbial communities. As a result, qCO_2 values are reduced if the MBC content increases and vice versa.

5. Conclusions

As far as we know, this is the first study focusing on the effects of land uses intensities on both bacterial communities and biological fertility index in soil cropped in Southern Italy, a typical Mediterranean region. The number of bacterial gene abundances was negatively regulated by soil microbial respiration. As indicated by qPCR and sequencing analyses, different land uses altered bacterial abundance and community structure; in particular, copiotrophic bacterial groups in the grass-covered vineyard decreased, while oligotrophic bacterial taxa increased along with a conservative managing of crop residues and no-tillage. In addition, the determination of SOM and microbial activity parameters allowed deriving relevant conclusions about the effect of land use intensities on the three studied ecosystems. The grass-covered vineyard demonstrated the highest amount of SOM and MBC, while respiration activity, qM, and CUE were higher in the cherry farm. The BFI indicated a higher soil biological fertility in the grass-covered vineyard, which can be reasonably assumed by the preservation of this ecosystem, with none or lower disturbance. These results could be useful to farmers in decisions about the best practices to avoid or reduce the negative impacts of land use intensities. Thus, the BFI could be an effective indicator system in the perceptions and assessment of soil fertility. More in-depth studies could validate the results of the present research under changing climate conditions and agro-ecosystems and allow the use of the biological fertility index to assess soil quality in agricultural soils.

Author Contributions: Conceptualization: M.Y.K. and C.C.; methodology: M.Y.K., P.M., and S.S.; software: M.Y.K.; validation: C.C.; formal analysis: M.Y.K. and C.C.; investigation: M.Y.K., P.M., and S.S.; resources: C.C.; data curation: M.Y.K. and C.C.; writing—original draft preparation: M.Y.K. and C.C.; writing—review and editing: C.C.; visualization: C.C.; supervision: C.C.; project administration: M.Y.K.; funding acquisition: M.Y.K. and C.C.

Funding: This research was funded by Global Doc project-CUP H96J17000160002- financed by Regione Puglia, Italy.

Acknowledgments: The authors thank Matteo Spagnuolo and Mr. Nicola Carella for scientific and technical assistance, respectively.

Conflicts of Interest: The authors declare no conflict of interest.

References

1. Cerdà, A.; Hooke, J.; Romero-Diaz, A.; Montanarella, L.; Lavee, H. Soil erosion on Mediterranean type-ecosystems. *Land Degrad. Dev.* **2010**, *21*, 71–74. [CrossRef]

2. Shelef, O.; Stavi, I.; Zdruli, P.; Rachmilevitch, S. Land use change, a case study from southern Italy: General implications for agricultural subsidy policies. *Land Degrad. Dev.* **2014**, *27*, 868–870. [CrossRef]

3. Zdruli, P.; Calabrese, J.; Ladisa, G.A.O. Impacts of land cover change on soil quality of manmade soils cultivated with table grapes in the Apulia Region of south-eastern Italy. *Catena* **2014**, *121*, 13–21. [CrossRef]

4. Bai, X.Y.; Wang, S.J.; Xiong, K.N. Assessing spatial-temporal evolution processes of karst rocky desertification land: Indications for restoration strategies. *Land Degrad. Dev.* **2013**, *24*, 47–56. [CrossRef]

5. Bajocco, S.; Smiraglia, D.; Scaglione, M.; Raparelli, E.; Salvati, L. Exploring the role of land degradation on agricultural land use change dynamics. *Sci. Total Environ.* **2018**, *636*, 1373–1381. [CrossRef]

6. Masciopinto, C.; De Giglio, O.; Scrascia, M.; Fortunato, F.; La Rosa, G.; Suffredini, E.; Pazzani, C.; Prato, R.; Montagna, M.T. Human health risk assessment for the occurrence of enteric viruses in drinking water from wells: Role of flood runoff injections. *Sci. Total Environ.* **2019**, *666*, 559–571. [CrossRef]

7. Zdruli, P. Land resources of the Mediterranean: Status, pressures, trends and impacts on future regional development. *Land Degrad. Dev.* **2014**, *25*, 373–384. [CrossRef]

8. Lindtner, P.; Gömöryová, E.; Gömöry, D.; Stašiov, S.; Kubovčík, V. Development of physico-chemical and biological soil properties on the European ground squirrel mounds. *Geoderma* **2019**, *339*, 85–93. [CrossRef]

9. Pepper, I.L.; Brusseau, M.L. Physical-chemical characteristics of soils and the subsurface. *Environ. Pollut. Sci.* **2019**, 9–22. [CrossRef]

10. Juhos, K.; Czigany, S.; Madarasz, B.; Ladanyi, M. Interpretation of soil quality indicators for land suitability assessment–A multivariate approach for Central European arable soils. *Ecol. Indic.* **2019**, *99*, 261–272. [CrossRef]

11. Li, P.; Shi, K.; Wang, Y.; Kong, D.; Liu, T.; Jiao, J.; Liu, M.; Li, H.; Hu, F. Soil quality assessment of wheat-maize cropping system with different productivities in China: Establishing a minimum data set. *Soil Tillage Res.* **2019**, *190*, 31–40. [CrossRef]

12. Francaviglia, R.; Renzi, G.; Ledda, L.; Benedetti, A. Organic carbon pools and soil biological fertility are affected by land use intensity in Mediterranean ecosystems of Sardinia, Italy. *Sci. Total Environ.* **2017**, *599–600*, 789–796. [CrossRef] [PubMed]

13. Renzi, G.; Canfora, L.; Salvati, L.; Benedetti, A. Validation of the soil Biological Fertility Index (BFI) using a multidimensional statistical approach: A country-scale exercise. *Catena* **2017**, *149*, 294–299. [CrossRef]

14. Laudicina, V.A.; Novara, A.; Barbera, V.; Egli, M.; Badalucco, L. Long-term tillage and cropping system effects on chemical and biochemical characteristics of soil organic matter in a Mediterranean semiarid environment. *Land Degrad. Dev.* **2015**, *26*, 45–53. [CrossRef]

15. Soleimani, A.; Hosseini, S.M.; MassahBavani, A.R.; Jafari, M.; Francaviglia, R. Influence of land use and land cover change on soil organic carbon and microbial activity in the forests of northern Iran. *Catena* **2019**, *177*, 227–237. [CrossRef]

16. Ahmed, I.U.; Mengistie, H.K.; Godbold, D.L.; Sanden, H. Soil moisture integrates the influence of land-use and season on soil microbial community composition in the Ethiopian highlands. *Appl. Soil Ecol.* **2019**, *135*, 85–90. [CrossRef]

17. Kome, G.; Enang, R.; Yerima, B.; Lontsi, M. Models relating soil pH measurements in H_2O, KCl and $CaCl_2$ for volcanic ash soils of Cameroon. *Geoderma Reg.* **2018**, *14*, e00185. [CrossRef]

18. Soil Survey Staff. *Soil Survey Laboratory Information Manual*; Soil Survey Investigations Report No. 45, Version 2.0.; USDA-Natural Resources Conservation Service: Washington, DC, USA, 2011.

19. Walkley, A.; Black, I.A. An examination of the Degtjareff method for determining soil organic matter, and a proposed modification of the chromic acid titration method. *Soil Sci.* **1934**, *37*, 29–38. [CrossRef]

20. Martínez, J.M.; Galantini, J.; Duval, M.; López, F.; Iglesias, J. Estimating soil organic carbon in Mollisols and its particle-size fractions by loss-on-ignition in the semiarid and semi humid Argentinean Pampas. *Geoderma Reg.* **2018**, *12*, 49–55. [CrossRef]

21. Vance, E.D.; Brookes, P.C.; Jenkinson, D.S. An extraction method for measuring soil microbial biomass C. *Soil Biol. Biochem.* **1987**, *9*, 703–707. [CrossRef]

22. Isermeyer, H. EineeinfacheMethodezurBestimmung der Bodenatmung und der KarbonateimBoden. *J. Plant. Nutr. Soil Sci.* **1952**, *56*, 26–38.

23. Liu, X.; RezaeiRashti, M.; Dougall, A.; Esfandbod, M.; van Zwieten, L.; Chen, C. Subsoil application of compost improved sugarcane yield through enhanced supply and cycling of soil labile organic carbon and nitrogen in an acidic soil at tropical Australia. *Soil Tillage Res.* **2018**, *180*, 73–81. [CrossRef]

24. Mganga, K.Z.; Razavi, B.S.; Kuzyakov, Y. Land use affects soil biochemical properties in Mt. Kilimanjaro region. *Catena* **2016**, *141*, 22–29. [CrossRef]

25. Geyer, K.; Dijkstra, P.; Sinsabaugh, R.; Frey, S. Clarifying the interpretation of carbon use efficiency in soil through methods comparison. *Soil Biol. Biochem.* **2019**, *128*, 79–88. [CrossRef]

26. Herlemann, D.P.R.; Labrenz, M.; Juergens, K.; Bertilsson, S.; Waniek, J.J.; Anderrson, A.F. Transition in bacterial communities along the 2000 km salinity gradient of the Baltic Sea. *ISME J.* **2011**, *5*, 1571–1579. [CrossRef] [PubMed]

27. Pascazio, S.; Crecchio, C.; Scagliola, M.; Mininni, A.N.; Dichio, B.; Xiloyannis, C.; Sofo, A. Microbial-based soil quality indicators in irrigated and rainfed soil portions of Mediterranean olive and peach orchards under sustainable management. *Agric. Water Manag.* **2018**, *195*, 172–179. [CrossRef]

28. Martin-Didonet, C.C.G.; Chubatsu, L.S.; Souza, E.M.; Kleina, M.; Rego, F.G.M.; Rigo, L.U.; Yates, M.G.; Pedrosa, F.O. Genome structure of the Genus *Azospirillum*. *J. Bacteriol.* **2000**, *182*, 4113–4116. [CrossRef]

29. Wallenstein, M.D.; Vilgalys, R.J. Quantitative analyses of nitrogen cycling genes in soils. *Pedobiologia* **2005**, *49*, 665–672. [CrossRef]

30. Kuczynski, J.; Stombaugh, J.; Walters, W.A.; González, A.; Caporaso, J.G.; Knight, R. Using QIIME to analyze 16S rRNA gene sequences from microbial communities. *Curr. Protoc. Biol.* **2011**, *27*. [CrossRef]

31. Jat, H.; Datta, A.; Choudhary, M.; Yadav, A.; Choudhary, V.; Sharma, P.; Gathala, M.; Jat, M.; McDonald, A. Effects of tillage, crop establishment and diversification on soil organic carbon, aggregation, aggregate associated carbon and productivity in cereal systems of semi-arid Northwest India. *Soil Tillage Res.* **2019**, *190*, 128–138. [CrossRef]

32. Qiu, Q.; Wu, L.; Li, B. Crop residue-derived dissolved organic matter accelerates the decomposition of native soil organic carbon in a temperate agricultural ecosystem. *Acta Ecol.Sinica.* **2019**, *39*, 69–76. [CrossRef]

33. Novara, A.; Minacapilli, M.; Santoro, A.; Rodrigo-Comino, J.; Carrubba, A.; Sarno, M.; Venezia, G.; Gristina, L. Real cover crops contribution to soil organic carbon sequestration in sloping vineyard. *Sci. Total Environ.* **2019**, *652*, 300–306. [CrossRef] [PubMed]

34. Ramesh, T.; Bolan, N.; Kirkham, M.; Wijesekara, H.; Kanchikerimath, M.; Srinivasa Rao, C.; Sandeep, S.; Rinklebe, J.; Ok, Y.; Choudhury, B.; et al. Soil organic carbon dynamics: Impact of land use changes and management practices: A review. *Adv. Agron.* **2019**. [CrossRef]

35. Safaei, M.; Bashari, H.; Mosaddeghi, M.; Jafari, R. Assessing the impacts of land use and land cover changes on soil functions using landscape function analysis and soil quality indicators in semi-arid natural ecosystems. *Catena* **2019**, *177*, 260–271. [CrossRef]

36. Fujita, K.; Miyabara, Y.; Kunito, T. Microbial biomass and ecoenzymaticstoichiometries vary in response to nutrient availability in an arable soil. *Eur. J. Soil Biol.* **2019**, *91*, 1–8. [CrossRef]

37. Denardin, L.; Carmona, F.; Veloso, M.; Martins, A.; Freitas, T.; Carlos, F.; Marcolin, É.; Camargo, F.; Anghinoni, I. No-tillage increases irrigated rice yield through soil quality improvement along time. *Soil Tillage Res.* **2019**, *186*, 64–69. [CrossRef]

38. Hao, M.; Hu, H.; Liu, Z.; Dong, Q.; Sun, K.; Feng, Y.; Li, G.; Ning, T. Shifts in microbial community and carbon sequestration in farmland soil under long-term conservation tillage and straw returning. *Appl. Soil Ecol.* **2019**, *136*, 43–54. [CrossRef]

39. González-Ubierna, S.; Lai, R. Modelling the effects of climate factors on soil respiration across Mediterranean ecosystems. *J. Arid Environ.* **2019**. [CrossRef]

40. Malik, A.A.; Puissant, J.; Buckeridge, K.M.; Goodall, T.; Jehmlich, N.; Chowdhuty, S.; Gweon, H.S.; Peyton, J.M.; Mason, K.E.; Agtmaal, M.V.; et al. Land use driven change in soil pH affects microbial carbon cycling processes. *Nat. Commun.* **2018**, *9*, 3591. [CrossRef]

41. Van Wesemael, B.; Chartin, C.; Wiesmeier, M.; von Lützow, M.; Hobley, E.; Carnol, M.; Krüger, I.; Campion, M.; Roisin, C.; Hennart, S.; et al. An indicator for organic matter dynamics in temperate agricultural soils. *Agric. Ecosyst. Environ.* **2019**, *274*, 62–75. [CrossRef]

42. Nannipieri, P.; Ascher, J.; Ceccherini, M.; Landi, L.; Pietramellara, G.; Renella, G. Microbial diversity and soil functions. *Eur.J. Soil Sci.* **2003**, *54*, 655–670. [CrossRef]

43. Nsabimana, D.; Haynes, R.J.; Wallis, F.M. Size, activity and catabolic diversity of the soil microbial biomass as affected by land use. *Appl. Soil Ecol.* **2004**, *26*, 81–92. [CrossRef]

44. Mocali, S.; Paffetti, D.; Emiliani, G.; Benedetti, A.; Fani, R. Diversity of heterotrophic aerobic cultivable microbial communities of soils treated with fumigants and dynamics of metabolic, microbial, and mineralization quotients. *Biol. Fertil. Soils.* **2008**, *44*, 557–569. [CrossRef]

45. Birge, H.E. What Happens During Soil Incubation? Exploring Microbial Biomass, Carbon AVAILABILITY and temperature Constraints on Soil Respiration. Master's Thesis, Colorado State University, Fort Collins, CO, USA, 2013.

46. Dungait, J.A.; Hopkins, D.W.; Gregory, A.S.; Whitmore, A.P. Soil organic matter turnover is governed by accessibility not recalcitrance. *Glob. Chang. Biol.* **2012**, *18*, 1781–1796. [CrossRef]

47. Basso, F.; Bove, E.; Dumontet, S.; Ferrara, A.; Pisante, M.; Quaranta, G.; Taberner, M. Evaluating environmental sensitivity at the basin scale through the use of geographic information systems and remote sensed data: An example covering the Agri basin (southern Italy). *Catena* **2000**, *40*, 19–35. [CrossRef]

48. Willy, D.; Muyanga, M.; Mbuvi, J.; Jayne, T. The effect of land use change on soil fertility parameters in densely populated areas of Kenya. *Geoderma* **2019**, *343*, 254–262. [CrossRef] [PubMed]

49. Ghimire, B.; Ghimire, R.; VanLeeuwen, D.; Mesbah, A. Cover crop residue amount and quality effects on soil organic carbon mineralization. *Sustainability* **2017**, *9*, 2316. [CrossRef]

50. Papini, R.; Valboa, G.; Favilli, F.; L'Abate, G. Influence of land use on organic carbon pool and chemical properties of VerticCambisols in central and southern Italy. *Agric. Ecosyst. Environ.* **2011**, *140*, 68–79. [CrossRef]

51. Wang, J.; Liu, G.; Zhang, C.; Wang, G.; Fang, L.; Cui, Y. Higher temporal turnover of soil fungi than bacteria during long-term secondary succession in a semiarid abandoned farmland. *Soil Tillage Res.* **2019**, *194*, 104305. [CrossRef]

52. Sainju, U.M.; Lenssen, A.W.; Allen, B.L.; Stevens, W.B.; Jabro, J.D. Nitrogen balance in response to dryland crop rotations and cultural practices. *Agric. Ecosyst. Environ.* **2016**, *233*, 25–32. [CrossRef]

53. Agnieszka, W.; Zofia, S.; Aleksandra, B.; Artur, B. Evaluation of factors influencing the biomass of soil microorganisms and DNA content. *Open J. Soil Sci.* **2012**, *2*, 64–69. [CrossRef]

54. Kallenbach, C.M.; Frey, S.D.; Grandy, A.S. Direct evidence for microbialderived soil organic matter formation and its ecophysiological controls. *Nat. Commun.* **2016**, *7*, 13630. [CrossRef] [PubMed]

55. Männistö, M.; Ganzert, L.; Tiirola, M.; Haggblom, M.M.; Stark, S. Do shifts in life strategies explain microbial community responses to increasing nitrogen in tundra soil? *Soil Biol. Biochem.* **2016**, *96*, 216–228. [CrossRef]

56. Fierer, N.; Bradford, M.A.; Jackson, R.B. Toward an ecological classification of soil bacteria. *Ecology* **2007**, *88*, 354–1364. [CrossRef]

57. Fontaine, S.; Mariotti, A.; Abbadie, L. The priming effect of organic matter: A question of microbial competition? *Soil Biol. Biochem.* **2003**, *35*, 837–843. [CrossRef]

58. Jiang, X.; Wright, A.L.; Wang, X.; Liang, F. Tillage-induced changes in fungal and bacterial biomass associated with soil aggregates: A long-term field study in a subtropical rice soil in China. *Appl. Soil Ecol.* **2011**, *48*, 168–173. [CrossRef]

59. Nazaries, L.; Tottey, W.; Robinson, L.; Khachane, A.; Abu-al-Soud, W.; Sorensen, S.; Singh, B.K. Shifts in the microbial community structure explain the response of soil respiration to land-use change but not to climate warming. *Soil Biol. Biochem.* **2015**, *89*, 123–134. [CrossRef]

60. Navarro-Noya, Y.E.; Gómez-Acata, S.; Montoya-Ciriaco, N.; Rojas-Valdez, A.; Suárez-Arriaga, M.C.; Valenzuela-Encinas, S.; Jiménez-Bueno, N.; Verhulst, N.; Govaerts, B.; Dendooven, L. Relative impacts of tillage, residue management and crop-rotation on soil bacterial communities in a semi-arid agroecosystem. *Soil Biol. Biochem.* **2013**, *65*, 86–95. [CrossRef]

61. Rincon-Florez, V.A.; Clement, N.G.; Dang, Y.; Schenk, P.M.; Carvalhais, L.C. Short-term impact of an occasional tillage on microbial communities in a Vertosol after 43 years of no-tillage or conventional tillage. *Eur. J. Soil Biol.* **2016**, *74*, 32–38. [CrossRef]

62. Li, J.; Wu, X.; Gebremikael, M.T.; Wu, H.; Cai, D. Response of soil organic carbon fractions, microbial community composition and carbon mineralization to high-input fertilizer practices under an intensive agricultural system. *PLoS ONE* **2018**, *13*, e0195144. [CrossRef]

Review

Diversity of Rhizobia and Importance of Their Interactions with Legume Trees for Feasibility and Sustainability of the Tropical Agrosystems

Emanoel G. Moura [1,*], Cristina S. Carvalho [1], Cassia P. C. Bucher [2], Juliana L. B. Souza [3], Alana C. F. Aguiar [4], Altamiro S. L. Ferraz Junior [1], Carlos A. Bucher [5] and Katia P. Coelho [6,*]

[1] Postgraduate Program in Agroecology, Maranhão State University, Lourenço Vieira da Silva Avenue, 1000, Jardim São Cristovão, São Luís 65055-310, MA, Brazil; carvalhoscristina@gmail.com (C.S.C.); altamiro.ferraz@gmail.com (A.S.L.F.J.)

[2] Department of Soils, Federal Rural University of Rio de Janeiro, BR 465, Km 07, Seropédica 23890-000, RJ, Brazil; cassiapcoelho04@gmail.com

[3] Postgraduate Program in Biodiversity and Conservation, Federal University of Maranhão, Portuguese Avenue, 1966, Bacanga, São Luís 65080-805, MA, Brazil; lbritos.juliana@gmail.com

[4] Biology Department, Federal University of Maranhão, Portuguese Avenue, 1966, Bacanga, São Luís 65080-805, MA, Brazil; alana.aguiar@ufma.br

[5] Department of Fitotecnia, Federal Rural University of Rio de Janeiro, BR 465, Km 07, Seropédica 23890-000, RJ, Brazil; carlos.bucher@gmail.com

[6] Agricultural Engineering Department/Postgraduate Program in Agroecology, Maranhão State University, Lourenço Vieira da Silva Avenue, 1000, Jardim São Cristovão, São Luís 65055-310, MA, Brazil

[*] Correspondence: egmoura@elointernet.com.br (E.G.M.); katiapc04@gmail.com (K.P.C.)

Received: 31 March 2020; Accepted: 30 April 2020; Published: 24 May 2020

Abstract: Symbiotic biological nitrogen fixation (BNF) is a complex process that involves rhizobia, a diverse group of α and β-proteobacteria bacteria, and legume species. Benefits provided by BNF associated with legume trees in tropical environments include improvements to efficiency of nitrogen (N) use, increase of soil carbon sequestration, stabilization of soil organic matter, decrease of soil penetration resistance, and improvement of soil fertility. All these benefits make BNF a crucial ecosystem service to the sustainability of tropical agriculture. Due to the importance of this ecological process and the high diversity of rhizobia, these bacteria have been extensively characterized worldwide. Currently, over 400 species of rhizobia are known, distributed into seven families. In the humid tropics, *Leucaena leucocephala*, *Acacia mangium*, *Gliricidia sepium*, and *Clitoria fairchildiana* are four of the most common species used by family farmers to create sustainable agricultural systems. These four legumes perform symbiosis with different groups of rhizobia. Exploring BNF could help to enable sustainable intensification of agriculture in the humid tropics, mainly because it can increase N use efficiency in an environment where N is a limiting factor to plant growth.

Keywords: biological nitrogen fixation; nitrogen; *Leucaena leucocephala*; *Acacia mangium*; *Gliricidia sepium*; *Clitoria fairchildiana*

---

1. Rhizobia Strains and Legume Tree Interactions: Importance and Use in Tropical Agricultural Systems

In the humid tropics, the edaphoclimatic conditions, which involve soils with fragile structure subjected to high temperatures, rainfall, and insolation, are unfavorable for organic matter accumulation and thus N availability and nitrogen use efficiency. The adoption of inappropriate agricultural practices for local conditions reduces nutrient availability and results in the depletion of soil fertility. Small family farmers in the humid tropics, especially in the pre-Amazon region, practice itinerant agriculture that is associated with the slashing and burning of natural vegetation. These small farmers mainly cultivate food crops such as rice, maize, cassava, and beans, using low technology. According to Moura et al. [1], this system has negative effects on the local and global environment and no longer provides social benefits to rural communities. The environmental impact of agriculture in tropical regions can include deforestation and loss of wildlife habitat, soil nutrient depletion, increased greenhouse gas emissions, and loss of biodiversity [2]. In contrast, there are excellent opportunities to avoid environmental damage through the use of ecosystem services to achieve ecological and economic benefits [3].

No-tillage alley cropping can be an alternative for the family farmers for maintaining productivity in the low-fertility soils of humid tropical regions. In this system, legume trees or shrubs are planted in two or more sets of single or multiple rows. Before and during the cropping period, the leguminous branches are periodically pruned and laid down on the soil surface between the leguminous tree sets, where other crops are planted. The use of legume trees can provide ecosystem services such as biomass production, recycling of nutrients, biological nitrogen fixation, and carbon sequestration [4].

Following the work of Fischer et al. [5], we define ecosystem services as the ecological processes of ecosystems that are used to produce human well-being. In this context, nitrogen (N) fixation by legume trees, in which bacteria living in root nodules convert atmospheric dinitrogen (N_2) gas to a plant-available form of N, is crucial to the sustainability of tropical agriculture [6]. Fixation of N in this manner can provide much of the N needed to drive crop growth and biomass creation [7]. In addition, proper management of legume tree biomass can increase soil organic matter, sequestered carbon, and recycled nutrients, which increases soil fertility and agricultural productivity [8]. In the tropics, the high biodiversity in rhizobia and legume trees, combined with long growing seasons and adequate soil moisture, provides the ideal conditions for making use of the ecosystem services provided by legume trees [9].

These phenomena offer a potential solution to agricultural sustainability challenges in the Amazon region, which are mainly associated with depletion of soil fertility [10]. The region's low-activity clay soils present difficulties for agriculture due to their reduced capacity to retain nutrients under high rainfall intensity, making agriculture an unfeasible activity [11]. Nutrient retention in the root zone can be enhanced by adding nutrients in slow-release forms and through biologically mediated processes like N-fixing symbiosis [12]. In the agroecosystems of the humid tropics, these approaches may be more sustainable than saturating the soil with soluble nutrients [13].

Biologically mediated processes are one of the most important targets in modifying and improving N uptake efficiency by roots [14]. This assertion is particularly true in humid tropical conditions, where N use efficiency is usually very low. High temperatures increase the volatilization rates of fertilizers [15], while high rainfall rates increase nitrate (NO^-_3) concentration and leaching [16]. Inefficient use of N, beyond its negative economic impact, contributes to agriculture disservices through greenhouse gas emissions and groundwater pollution [17]. Indeed, in the humid tropics, N uptake is impaired by N leaching due to high rainfalls, and by reduced root growth in hard-setting soil when rainfall is low [18]. The poor agronomic efficiency of N, reaching as low as 14 kg/kg (maize grain/applied N) has been the main reason why many family farmers have resisted changing from the traditional slash-and-burn systems to conventional tillage systems, despite recommendations to switch.

Some authors have reported the positive effect of the symbiosis between rhizobia and legume trees on agroecosystem services, one of the main results of which is an increase in tropical agriculture sustainability. Sena et al. [10] showed that the total N content of maize was increased by around 50% when adding biomass of gliricidia (*Gliricidia sepium* (Jacq.)) on its own and with urea. Aguiar et al. [13] reported that a combination of high- and low-quality residues from legume trees increased N concentrations in maize. A combination of gliricidia with sombreiro (*Clitoria fairchildiana* (R. A. Howard)) resulted in a nearly threefold increase in N accumulation compared to bare soil with urea (148.4 vs 54.1 kg ha^{-1}). Treatments involving legume residues plus urea led to higher N accumulation in comparison to the treatments with residues but without urea [13].

N fixation by legume trees also has a positive effect on soil carbon sequestration. N-fixing trees sequester CO_2 directly through their growth or indirectly via the turnover of their N-rich tissues, whose decomposition increases plant growth due to higher soil N [7]. However, some authors have highlighted the influence of quality of the applied residue (i.e., residues with high N contents and low to medium cellulose contents) on the stabilization of the soil organic matter [19,20]. According to Bradford et al. [21], high-quality leaf litter, which increases microbial activity, results in more organic matter that can be physicochemically stabilized. Therefore, the higher the symbiotic efficiency between rhizobia and legume trees, the higher the quality and amount of stabilized organic matter or sequestered carbon. According to Sena et al. [10], total carbon stock was 30% higher in treatments with high-quality legume tree residue than in treatments with bare soil. Thus, compared with the control treatment, the treatment with gliricidia, when added to the accumulated organic content, resulted in the carbon stock increasing by 5.0 g kg^{-1}.

Improvement in soil fertility from legume tree biomass can be physical (decreased soil penetration resistance), or chemical (increased base cations due to recycled nutrients) [18]. After four rainless days, the effect of biomass application on penetration resistance in a hard-setting soil extended to a depth of 17.5 cm, due less to soil moisture conservation than to the increase in soil organic matter [22]. This biomass may also capture a considerable amount of recycled base cations that were previously out of the reach of crops, improving the environment of the root zone [9,23].

Since the Forest Code in Brazil specifies that only 20% of a farmer's land can be used for agriculture in the Amazon, local research and policy agendas must consider N-fixing symbiosis between rhizobia and legume trees. This would allow sustainable intensification of agriculture in this region, for which increased N availability and use efficiency is crucial [24]. However, strategy to meet the challenges of the systematic use of ecosystem services provided by interactions of leguminous trees and rhizobia will require the mastery of a vast knowledge base [25]. First, we need to know how symbiotic relationships between different species could work through the ecological gradient, like the one that makes up the Amazon and its periphery. Such a process can determine the efficiency of N input in tropical family farm systems [26], increasing the ease of adoption by farmers. Second, it will be necessary to identify and select rhizobia communities which have greater efficiency to produce high quality biomass if the aim is to take advantage of the large tropical biodiversity [24].

A large number of texts have recently appeared in the scientific literature on these different issues, but often with an orientation less applied to the tropical environment and its agrosystems. This paper provides an overview of recent developments in the diversity of rhizobia for N fixation but is mostly concerned with contributing to meet the challenges of feasibility and sustainability of the agrosystems in tropical family farms.

2. Biological N Fixation and Indigenous Rhizobia Communities in Tropical Environments

N is the most limiting nutrient for plants, especially in the tropics, where it is particularly important in secondary forest succession [7], which is common in the humid tropics due to the practice of shifting cultivation. The largest source of N is the atmosphere, where it is found as N_2, a form that is unavailable to most living beings. Most plant-available N is extracted from the atmosphere using biological nitrogen fixation (BNF).

N can also be fixed by the Haber–Bosch process, an industrial procedure used to produce N fertilizers, or by electrical discharges in the atmosphere. However, the Haber–Bosch process has a high economic and environmental cost, since it requires a high amount of energy, requires fossil fuel inputs, and generates harmful by-products [27]. BNF is important in agriculture due to its low/no cost and non-polluting process. This environmental service contributes to the production of high protein content seeds for food and forage. In addition, it considerably reduces the use of fertilizers, favoring the productivity of ecosystems and reducing losses from N leaching [28]. BNF consists of the conversion of atmospheric N to ammonia and can be carried out by free-living, associative, or symbiotic bacteria. This symbiosis represents an inexpensive and sustainable approach to crop production [27]. BNF is mediated by nitrogenase, a complex metalloenzyme, with well-preserved structural and mechanical characteristics.

Nodulation in legumes evolves a highly specific interaction between these plants and rhizobia, which are gram-negative soil bacteria. This symbiosis includes several processes, including the recognition of symbionts, infection and colonization of plants by bacteria, the formation of root nodules by the plant, and biological N fixation. Modern genetic and molecular methods are effective in identifying N-fixing organisms [29]. Indigenous or native rhizobia are those naturally found in the soils of a given location. Studies have shown that tropical soils have native rhizobia communities with high diversity and varied efficiency and are a source of genetic variability in the search for strains capable of being used as inoculants [25,30,31]. Inoculants are products composed of living microorganisms capable of benefiting the development of different plant species. The rhizobia were the first microorganisms used as inoculants. Currently, the use of inoculants is widespread and indicated in agriculture, mainly for legumes such as soybeans, common beans, faba beans, and cowpea, but the production of inoculants for other legumes and non-legumes has increased in order to obtain greater yield [32].

The high variability among strains of indigenous communities is largely the result of the numerous stresses and natural interactions to which these communities are subject due to edaphoclimatic factors. However, it must be emphasized that the genes that regulate BNF are highly conserved in the organisms that express it. Therefore, they are not subject to significant variations [29,33]. The efficiency of strains in the native community has a direct impact on BNF with legumes that are native or introduced in certain areas. BNF is determined by the efficiency and number of rhizobia present in the soil and influenced by the host and environmental conditions [34,35]. The characteristics of the native community of rhizobia, especially features related to competitiveness and efficiency, are critical for the management of legumes of agricultural importance or those that want to explore the BNF. Inoculation with efficient rhizobia is usually recommended for locations where there are no compatible rhizobia, reduced soil rhizobia populations, or where the native community of rhizobia has low efficiency [36].

Currently, the benefits of using inoculants are well known and the target of frequent studies. However, potential environmental impacts related to this practice are generally overlooked. The use of microbial inoculants can influence the indigenous microbial communities. In addition, there is a major concern about how the impact on taxonomic groups can be related to effects on functional capabilities of the soil microbial communities. The use of microbial inoculants may cause major changes in microbial soil communities. This change is capable of ultimately inducing unpredictable feedback reactions. This is because the responses in vegetative development (plant growth), contrary to expectations, may be related to indirect effects of using the inoculant such as induction or repression of resident microbial populations which can influence beneficial soil functions such as biological nitrogen fixation [37].

Bakhoum et al. [38] demonstrated that, in addition to the inoculant, soil origin and the plant provenance influenced the plant growth, bacterial structure, and diversity and soil functioning in plant rhizosphere. This means that inoculation with selected rhizobial strains is a suitable tool for increasing plant growth; however, for each soil origin, plant species, or provenance, a rhizobial strain must be objectively selected. In this scenario, an inoculum mixture is an alternative to be considered. Thilakarathna and Raizada [39], through a meta-analysis that encompassed several studies on soybean inoculation, demonstrated that inoculants consisting of indigenous rhizobia, sometimes, result in greater yields than non-local, improved rhizobia under field conditions and that these native rhizobia are generally more competitive than the improved rhizobia in terms of nodule occupancy. This indicates that native rhizobia may have potential for commercial inoculant production at a local level. Additionally, the review pointed out that locally adapted rhizobia strains are capable of performing better under environmental stress conditions than introduced rhizobia. This characteristic is especially important for the choice of inoculant strains for some regions with limiting characteristics such as semi-arid regions.

The rhizobia host specificity refers to the ability to interact between symbionts. It is a concept that can be applied to both bacteria and plants. Legume root exudates contain phenolic compounds called flavonoids that may or may not activate nod gene expression in the bacterial partner. If activated, nod genes expression results in the production and secretion of a set of molecular signals known as nod factors, that may or may not elicit the appropriate responses in the plant required for rhizobial root infection and nodule development [40]. Strains of rhizobia and legumes that can establish symbiotic relationships with several species are considered promiscuous. Usually, a particular species of rhizobia can nodulate a limited rate of legumes [36,41,42]. Therefore, the harnessing of N fixation as a tool for agricultural development is dependent on immersing oneself in the vast diversity of species and characteristics to identify and select the agronomically more efficient ones. The symbiotic relationships between two species and the diversification of rhizobia in functional, adaptive characteristics through the ecological gradient can determine the N input in several ecological niches [36,43].

Biological nitrogen fixation includes several processes, such as the recognition of symbionts, infection and colonization of plants by bacteria, the formation of root nodules by the plant, and biological N fixation. The plants regulate the formation of root nodules that house nitrogen-fixing rhizobia and adjust investment into nodule development and growth [44]. Legumes occupy a variety of habitat types, and they can also take on a range of growth forms (trees, vines, shrubs, and herbs) and substantially differ in their promiscuity with rhizobial partners (generalists vs. specialists) and their ability to regulate symbiotic nitrogen fixation (obligate vs. facultative) [45].

For nitrogen-fixing rhizobia, assemblages are mostly determined by filtering by the host as well as abiotic soil conditions [46]. Therefore, legume–rhizobia association is a powerful model of the limits of host control over microbes [44].

An important consideration when assessing the effects of mutualisms such as legumes and rhizobia is the specificity/promiscuity of each partner in the symbiosis. If a species in the mutualism is a generalist (will engage in symbiosis with multiple partner species), then increasing partner diversity can increase its contribution to ecosystem functioning [45]. However, for legumes adapted to highly variable environments and nutrient-poor soils, filtering out rhizobial partners may not be an effective strategy to ensure symbiotic benefits [46]. Strain-specific legume rhizobia symbioses can develop in particular habitats and lateral gene transfer of specific symbiosis genes within rhizobial genera is an important mechanism allowing legumes to form symbioses with rhizobia adapted to particular soils [47]. The generalists (i.e., promiscuous legumes) obtain fewer average benefits from rhizobia than specialists [45]. However, the "Jack-of-all-trades is a master of none" hypothesis asserts that specialists persist because the fitness of a generalist utilizing a particular habitat is lower than that of a specialist adapted to that habitat [48].

Specialized interactions help structure communities, but persistence of specialized organisms is puzzling because a generalist can occupy more environments and partake in more beneficial interactions [48]. A specialist–generalist trade-off would suggest that hosts often benefit from blocking many rhizobia strains, which conflicts with the fitness interest of rhizobia to increase nodulation [44]. Therefore, species-specific legume–rhizobial interactions can directly influence niche and fitness differences among legume species and between legumes and non-legumes, with important consequences for the richness–productivity relationship [45].

Both the rhizobia–host specificity and the diversity of the native rhizobia community are fundamental for the efficient use of inoculants. Poor inoculation responses may be related to large and diverse indigenous populations of well adapted nodulating bacteria. This occurs with the inoculation of beans in Latin American agricultural fields [49].

3. Identification of Bacterial Strains and Determination of Rhizobia Diversity

Rhizobia have been extensively characterized in global studies due to their high diversity and importance for agriculture [50]. These studies of diversity generally involve sampling of nodules in the fields (soils and seeds), rhizobia isolation from nodules, identification of the isolates at the genus and species levels, molecular characterization through genotyping of isolates, and characterization of symbiosis-related genes, as well as nodulation tests, phenotyping, and description of new rhizobium species [51].

Currently, several molecular tools are available to study and identify bacterial diversity. Molecular methods have been developed to reveal the rhizobia diversity at the genetic, strain, species, genus, or higher levels [51]. Genetic analysis includes DNA–DNA hybridization, G + C contents, and PCR analysis using a large number of genes, including housekeeping genes [52]. For molecular characterization of the rhizobial strains, different PCR methods can be used [50]. To further improve the identification process, the analysis of conserved housekeeping genes has been used successfully to detect rhizobia diversity with precision [52]. The housekeeping genes used in PCR and sequence analyses for further characterization of representative strains of each *recA* genotype include the 16S rRNA, *atpD*, *glnII*, *dnaK*, *gap*, *glnA*, *gltA*, *gyrB*, *pnp*, *rpoB*, and *thrC*. Recommended strategies for studies on rhizobia diversity include 1, screening by *recA* phylogeny; 2, phylogenetic analyses of housekeeping genes; 3, BOX-PCR; 4, phenotypic characterization; 5, chemical taxonomy; 6, phylogenetic analyses of symbiosis genes and symbiotic specificity; 7, genome analysis; 8, description of new species and genera [51].

Sequencing the 16S rRNA gene, one of the most important tools in current studies on microbiology, has been used mainly to identify and classify isolates from pure cultures and estimate bacterial diversity in the environment without culture samples, through metagenomic approaches [53]. The 16S rRNA gene is a component of the 30S subunit of the prokaryote ribosomes, whose other components are 23S and 5S in the RNA molecules. The 16S gene is considered a good phylogenetic marker because it displays some conserved regions that are useful for unveiling phylogenetic relationships between distant species and also more variable regions, which are used to differentiate closely related species. For example, Salvina et al. [54] carried out preliminary tests to select the best procedure for genetic identification of rhizobia species and selected the 16S rRNA gene due to its reliable genetic identification of bacteria. Moreover, Hakim et al. [55] used the sequencing of the 16s rRNA by the illumina®system (Illumina, Inc, San Diego, USA) to obtain an overview of microbial diversity using a metagenomic approach and the relative distribution of the endophytic population (rhizobial and non-rhizobial) in mung bean nodules grown in different areas.

Although important, molecular methods should not be limited to the 16S rRNA gene alone, since a taxonomic consensus is best achieved when different types of data are combined. Such an integrated model, which includes phenotypic, genotypic, and phylogenetic information, is referred to as polyphasic taxonomy [56]. To identify rhizobia at the species and subspecies levels, identification and classification of bacteria and particularly rhizobia using polyphasic approaches is becoming the most accepted method [52].

Currently, several methodologies and other housekeeping genes have been used in phylogenetic studies of bacteria, mainly because the 16S rRNA has limitations in resolving the taxonomy and phylogeny of some genera, such as *Bradyrhizobium*, where high conservation of this gene hampers diversity analysis and species identification [56,57]. Other methods that have been used include characterization and screening of nodulating bacteria and sequencing of repetitive DNA-like BOX, ERIC sequences, and the 16S-23S rRNA intergenic spacer. However, so far, limited databases are available to compare results from different studies [52]. Moreover, screening by *recA* is advantageous because such phylogenetic analyses can simultaneously determine the genus and species of the rhizobial strains. On the other hand, many rhizobial species share highly similar (>97%) or even identical sequence of 16S rRNA.

A phenotypic or morphocultural characterization is a culture-dependent, fast, and low-cost method that provides information on groups of viable and cultivable microorganisms in a soil sample [58]. The phenotypic data involves biochemical and physiological traits, and analysis includes Gram staining; cell morphology and motility; oxidase and catalase activity; Biolog tests; NaCl tolerance, antibiotic, pH and temperature profiling; fatty acid composition; nodulation; and N fixation [52]. In this method, evaluating the morphological characteristics is the first step to identify new taxonomic groups of microorganisms. Due to its low cost, the phenotypic characterization can be useful in laboratories that do not have access to sophisticated technologies [59]. Currently, despite the high value of molecular methods to bacteria taxonomy, phenotypic characterization still plays an important role in classification [60]. The isolation and characterization of rhizobia from nodules collected in the field or from bait plants require few materials and are carried out, according to Vincent [61], in the following stages: 1, rehydration of nodules, if desiccated, in sterile water; 2, superficial disinfestation of nodules; 3, crushing and streaking the internal content of the nodule in a petri dish containing Yeast-extract-mannitol (YEM) broth. The most important characteristics observed are the time of emergence of isolated colonies and pH reaction in YEM broth with bromothymol blue; however, exopolysaccharide production, shape, size, color, texture, consistency, and optical details of the colony can also be observed [62]. Thomas-Oates et al. [63] also point out physiological and symbiotic characteristics as important for the characterization of strains of rhizobia.

Analyses of the chemical composition of *Rhizobium* cells include the content of cellular fatty acids, protein, respiratory quinones, polar lipids, and G + C mol% of the genomic DNA [53]. Although the data from these analyses are not so valuable for species differentiation, they can be used as descriptive characteristics for the species that are currently known [52].

Genome analysis complements studies on bacterial diversity. With the development of genome sequence analysis, the DNA–DNA hybridization (DDH) methods were replaced by average nucleotide identity (ANI) and digital hybridization of genome sequences in the description of novel species and genera [64,65]. Several analyses mentioned here are used to describe new species and genera. The International Committee on Systematics of Prokaryotes hosts the Subcommittee for the Taxonomy of Rhizobia and Agrobacteria, which holds regular meetings to discuss relevant issues and keep track of newly published species and genera. Importantly, the subcommittee publishes recommendations for the description of new species and genera of rhizobia and agrobacteria, and authors are expected to follow these guidelines [51].

Using these described methods, we identified different types of nodular bacteria and verified that nitrogen-fixing Leguminosae-nodulating bactéria (NFLNB) or rhizobia are included in a diverse phylogenetic classification list with species distributed in alphaproteobacteria and betaproteobacteria of the following genera (Table 1):

Table 1. Classification and number of species of bacteria identified as rhizobia.

Family	Genus/Type Species	Species of Rhizobia within the Genus, According to List of Prokaryotic Names with Standing in Nomenclature (2018)	References
	Rhizobium. Type species: *R. leguminosarum*	133 species	[66]
Rhizobiaceae	*Ensifer/Sinorhizobium* Type species: *E. adhaerens*; *S. fredii*; *S. xinjiangensis.*	The genus *Ensifer/Sinorhizobium* covers about 24 species	[67,68]
	Allorhizobium. Type species: *A. undicola*	9 species	[69]
	Pararhizobium. Type species: *P. giardinii.*	*P. giardinii*; *P. herbae*	[70]
	Shinella. Type species: *S. granuli*	*S. kummerowiae*	[71,72]
	Neorhizobium. Type species: *N. galegae*	*N. alkalisoli*, *N. galegae* and *N. huautlense*	[73]
Hyphomicrobiaceae	*Azorhizobium*. Type species: *A. caulinodans*	*A. caulinodans*, *A. doebereinerae* and *A. oxalatiphilum*	[74–76]
	Devosia. Type species: *D. riboflavina*	*D. neptuniae*	[77,78]
Bradyrhizobiaceae	*Bradyrhizobium* Type species: *B. japonicum*	60 species	[79–81]
	Blastobacter. Type species: *B. tienricii*	*B. denitrificans*	[82]

Table 1. *Cont.*

Family	Genus/Type Species	Species of Rhizobia within the Genus, According to List of Prokaryotic Names with Standing in Nomenclature (2018)	References
Phylobacteriaceae	*Mesorhizobium.* Type species: *M. loti*	59 species	[83,84]
	Aminobacter. Type species: *A. aminovorans*	*A. anthyllidis*	[85,86]
	Phyllobacterium Type species: *P. myrsinacearum*	8 species: *P. ifriqiyense, P. leguminum, P. bourgognense, P. brassicacearum, P. endophyticum, P. loti, P. sophorae, P. trifolii*	[87,88]
Methylobacteriaceae	*Methylobacterium.* Type species: *M. organophilum*	*M. nodulans*	[89,90]
	Microvirga Type species: *M. subterrânea*	*M. lupini, M. lotonononidis, M. zambiensis* and *M. vignae*	[91,92]
Brucellaceae	*Ochrobactrum.* Type species: *O. anthropi*	*O. lupini,* (synonym of *O. anthropic*); *O. cytisi*	[93–96]
Burkholderiaceae	*Burkholderia.* Type species: *B. cepacian.* Sawana et al. (2015) proposed the division of the genus into *Burkholderia* and *Paraburkholderia,*	The genus *Burkholderia* has 32 species and *Paraburkholderia* has 69 species	[97–99]
	Cupriavidus. Type species: *C. necator*	*C. taiwanenses*	[100–102]
	Ralstonia. Type species: *R. pickettii,*	*R. taiwanensis* (Sinonimous: *Cupriavidus taiwanensis; Wautersia taiwanensis*)	[103,104]

4. Diversity of Rhizobia in Nodules of Legume Trees in the Humid Tropics

In the humid tropics, more specifically, in the periphery of the Amazon region, legume trees are important for their ability to colonize and supply N in environments with poor soils due to the continuous production of biomass throughout the year, even in long periods of drought, which provides better soil coverage [105].

Legume trees of economic and environmental interest are distributed throughout the world, and some non-native species are widely distributed in the tropics [106]. These legumes can be used to recover degraded areas by providing N-rich biomass for agroforestry systems, produce high-quality forage for livestock, stabilize slopes against erosion, and provide shade for planting crops, fruits, and vegetables for human consumption, among others [33,107]. It should also be noted that the success of the BNF of legume, both introduced and native, depends on the presence of compatible and efficient rhizobia in the native community of the soil [106]. Therefore, the evaluation of the presence or absence of efficient strains in the native community of the area and forecast of the need for inoculation is the first step towards the establishment of legume trees, regardless of the purpose. This review will focus on four legume trees, the leucaena (*Leucaena leucocephala* (Lam.) de Wit), acacia (*Acacia mangium* Willd.), sombreiro (*C. fairchildiana* R. A. Howard), and gliricidia (*G. sepium* (Jacq.) Kunth), common in sustainable agricultural systems in the humid tropics due to their effectiveness in the use of N, mainly by low-income family farmers [10,13].

Leucaena is a fast-growing legume tree belonging to the subfamily Mimosoideae, native to Central America and widely distributed in tropical regions. It has a high BNF (average of 177–247 kg ha^{-1}), high leaf N content (about 1.3%), and high biomass production (about 5 t ha^{-1}) [108]. Leucaena is used to rehabilitate degraded areas by vegetation covering, in agroforestry systems, as forage, human food, firewood, wood, green manure, shade, support for scandent species, and to control wind and soil erosion [109]. It appears that leucaena is non-specific with respect to its symbiotic relationships with NFLNB, since it can form nodules with species of several genera of rhizobia. A survey of several studies carried out in tropical regions indicated that nodulation of leucaena occurs, preferably, with fast-growing rhizobia. Four recent studies have identified species of *Rhizobium* in leucaena nodules [105,109–111], while others have found species belonging to the genus *Mesorhizobium* [109–112]. Only one study reported the occurrence of symbiosis between leucaena and *Bradyrhizobium* [111] and between leucaena and *Cupriavidus* [113].

The BNF capacity of acacia in symbiosis with rhizobia is one of its most advantageous characteristics. This legume establishes symbiosis with species of *Bradyrhizobium*, *Rhizobium*, *Mesorhizobium*, and *Ochrobactrum* [114,115]. However, symbiosis between acacia and strains of *Bradyrhizobium* is more common than with species of other genera of rhizobia, as shown in studies on the identification of rhizobia strains in symbiosis with acacia in several regions [94,116–119]. Some authors have reported that effective nodulation of acacia occurs only with specific strains of *Bradyrhizobium* sp., such as *B. elkanii*, although acacia seedlings show high variability in their ability to fix N_2 in symbiosis with their specific strains of *Bradyrhizobium* [114,117,118,120].

Another legume of interest is gliricidia, a medium-sized tree that can reach up to 12 m in height [121]. The species is native to Mesoamerica and is considered the second most important multifunctional legume tree only behind leucaena [122]. Due to the high capacity of climate adaptation, this legume has been transported to most tropical countries, where it is widely distributed. Gliricidia can produce about 5 t ha^{-1} of dry matter and 186 kg ha^{-1} of N. The capacity for BNF makes the use of this species advantageous, especially for green manure purposes and for use in agroforestry systems. This legume establishes symbiosis, preferably with fast-growing rhizobia, mainly of the genus *Rhizobium* [25,123]. This fact is confirmed by the massive presence of species of the genus *Rhizobium* in a survey carried out with the main studies that addressed symbiosis of gliricidia with rhizobia, identified by a wide variety of molecular or classic methodologies [25,105,109,110,124,125].

The sombreiro is a rustic, medium to large fast-growing legume tree. The species is native to Brazil, and its phytogeographic domain includes the Amazon rainforest. Although it has a clear preference for fertile and moist soils, the sombreiro can also occur in open and altered areas [126]. The BNF capacity of this tree is known, but little studied, despite its high production of biomass (about 8 t ha^{-1}) and N (328 kg ha^{-1}) [127], with reports of symbiosis with *Rhizobium* strains [114] and *Bradyrhizobium* [25,120].

5. Final Considerations

The huge diversity of rhizobia strains and their interactions with legume trees in tropical soils, if used wisely, can contribute to the sustainability of tropical agroecosystems. According to studies carried out in these regions, tropical soils have a high diversity of rhizobia, which interact with different legumes. Furthermore, the efficiency of this symbiosis depends on both the symbionts and external factors. Indeed, BNF is a highly complex process that involves a diverse group of bacteria, currently distributed into seven families and 19 genera identified through several methods based on phenotypic/morphological, biochemical, and molecular characterizations, the latter of which is making major contributions to modern taxonomy. Fortunately, four of the most suitable legumes for use in achieving more sustainable agricultural systems for family farmers in the humid tropics, *L. leucocephala*, *A. mangium*, *G. sepium*, and *C. fairchildina*, perform symbiosis with different groups of rhizobia. Future research must be focused on efficiency of the interaction between the symbionts and external factors, which can lead to higher legume biomass, rich in N. In this scenario, the exploration of BNF as a key ecological service could bring economic, ecological, and agronomic benefits to assist in the process of sustainable intensification of agriculture in the humid tropics, mainly because BNF can increase the efficiency of N use in this environment, where it is a major limiting factor.

Author Contributions: Writing—original draft preparation, E.G.M. and K.P.C.; writing—review and editing, C.S.C.; C.P.C.B.; J.L.B.S.; A.C.F.A.; A.S.L.F.J., and C.A.B.; visualization, K.P.C. All authors have read and agreed to the published version of the manuscript.

Funding: This research was funded by "NUCLEUS, a virtual joint center to deliver enhanced N-use efficiency via an integrated soil–plant systems approach for the United Kingdom and Brazil", grant number: FAPESP—São Paulo Research Foundation [grant number 2015/50305-8]; FAPEG—Goiás Research Foundation [grant number 2015-10267001479]; and FAPEMA—Maranhão Research Foundation [grant number RCUK-02771/16]; and in the United Kingdom by the Biotechnology and Biological Sciences Research Council [grant number BB/N013201/1] under the Newton Fund scheme.

Acknowledgments: The authors thanks the National Council for Scientific and Technological Development (CNPq), the Foundation for the Support of Research and Scientific and Technological Development of Maranhão

(FAPEMA), and the Coordination for the Improvement of Higher Education Personnel (CAPES) by research grants from team members.

Conflicts of Interest: The authors declare no conflict of interest.

References

1. Moura, E.G.; Marques, E.S.; Silva, T.M.B.; Piedade, A.; Aguiar, A.C.F. Interactions among leguminous trees, crops and weeds in a no-till alley cropping system. *Int. J. Plant Prod.* **2014**, *8*, 441–456.
2. Power, A.G. Ecosystem services and agriculture: Tradeoffs and synergies. *Philos. Trans. R. Soc. B Biol. Sci.* **2010**, *365*, 2959–2971. [CrossRef] [PubMed]
3. Prado, R.B.; Fidalgo, E.C.C.; Monteiro, J.M.G.; Schuler, A.E.; Vezzani, F.M.; Garcia, J.R.; Oliveira, A.P.; Viana, J.H.M.; Gomes, B.C.C.P.; Mendes, I.C.; et al. Current overview and potential applications of the soil ecosystem services approach in Brazil. *Pesqui. Agropecu. Bras.* **2016**, *51*, 1021–1038. [CrossRef]
4. Medinski, T.; Freese, D. Soil carbon stabilization and turnover at alley-cropping systems, Eastern Germany. *Geophys. Res. Abst.* **2012**, *14*, 2012–9532.
5. Fisher, B.; Turner, R.K.; Morling, P. Defining and classifying ecosystem services for decision making. *Ecol. Econ.* **2009**, *68*, 643–653. [CrossRef]
6. Gehring, C.; Vlek, P.L.G.; De Souza, L.A.G.; Denich, M. Biological nitrogen fixation in secondary regrowth and mature rainforest of central Amazonia. *Agric. Ecosyst. Environ.* **2005**, *111*, 237–252. [CrossRef]
7. Batterman, S.A.; Hedin, L.O.; Van Breugel, M.; Ransijn, J.; Craven, D.J.; Hall, J.S. Key role of symbiotic dinitrogen fixation in tropical forest secondary succession. *Nature* **2013**, *502*, 224–227. [CrossRef] [PubMed]
8. Fernández, R.; Frasier, I.; Noellemeyer, E.; Quiroga, A. Soil quality and productivity under zero tillage and grazing on Mollisols in Argentina – A long-term study. *Geoderma Reg.* **2017**, *11*, 44–52. [CrossRef]
9. Berenguer, E.; Gardner, T.A.; Ferreira, J.; Aragão, L.E.O.C.; Nally, R.M.; Thomson, J.R.; Vieira, I.C.G.; Barlow, J. Seeing the woods through the aplings: Using wood density to assess the recovery of human-modified Amazonian forests. *J. Ecol.* **2018**, *106*, 2190–2203. [CrossRef]
10. Sena, V.G.L.; Moura, E.G.; Macedo, V.R.A.; Aguiar, A.C.F.; Price, A.H.; Mooney, S.J.; Calonego, J.C. Ecosystem services for intensification of agriculture, with emphasis on increased nitrogen ecological use efficiency. *Ecosphere* **2020**, *11*, 1–14. [CrossRef]
11. Glaser, B.; Lehmann, J.; Zech, W. Ameliorating physical and chemical properties of highly weathered soils in the tropics with charcoal - A review. *Biol. Fertil. Soils* **2002**, *35*, 219–230. [CrossRef]
12. Drinkwater, L.E.; Snapp, S.S. Nutrients in agroecosystems: Rethinking the management paradigm. *Adv. Agron.* **2007**, *92*, 163–186.
13. Aguiar, A.C.F.; Elialdo, S.A.; Anagila, C.S.J.; Moura, E.G. How leguminous biomass can increase yield and quality of maize grain in tropical agrosystems. *Legum. Res.* **2019**. [CrossRef]
14. Chapman, N.; Miller, A.J.; Lindsey, K.; Whalley, W.R. Roots, water, and nutrient acquisition: Let's get physical. *Trends Plant Sci.* **2012**, *12*, 701–710. [CrossRef] [PubMed]
15. Viero, F.; Bayer, C.; Fontoura, S.M.V.; Moraes, R.P. Ammonia volatilization from nitrogen fertilizers in no-till wheat and maize in southern. *Rev. Bras. Ciênc. Solo* **2014**, *38*, 1515–1525. [CrossRef]
16. Jabloun, M.; Schelde, K.; Tao, F.; Olesen, J.E. Effect of temperature and precipitation on nitrate leaching from organic cereal cropping systems in Denmark. *Eur. J. Agron.* **2015**, *62*, 55–64. [CrossRef]
17. Rutting, T.; Aronsson, H.; Delin, S. Efficient use of nitrogen in agriculture. *Nutr. Cycl. Agroecosyst.* **2018**, *110*, 1–5. [CrossRef]
18. Moura, E.G.; Portela, S.B.; Macedo, V.R.A.; Sena, V.G.L.; Souza, C.C.M.; Aguiar, A.C.F. Gypsum and legume residue as a strategy to improve soil conditions in sustainability of agrosystems of the humid tropics. *Sustainability* **2018**, *10*, 1006. [CrossRef]
19. Moura, E.G.; Sena, V.G.; Corrêa, M.S.; Aguiar, A.C.F. The importance of an alternative for sustainability of agriculture around the periphery of the Amazon rainforest. *Recent Pat. Food Nutr. Agric.* **2013**, *5*, 70–78. [CrossRef]
20. Martens, D.A. Plant residue biochemistry regulates soil carbon cycling and carbon sequestration. *Soil Biol. Biochem.* **2000**, *32*, 361–369. [CrossRef]

21. Bradford, M.A.; Keiser, A.D.; Davies, C.A.; Mersmann, C.A.; Strickland, M.S. Empirical evidence that soil carbon formation from plant inputs is positively related to microbial growth. *Biogeochemistry* **2013**, *113*, 271–281. [CrossRef]

22. Castellano, M.J.; Mueller, K.E.; Olk, D.C.; Sawyer, J.E.; Six, J. Integrating plant litter quality, soil organic matter stabilization, and the carbon saturation concept. *Glob. Chang. Biol.* **2015**, *21*, 3200–3209. [CrossRef] [PubMed]

23. Moura, E.G.; Oliveira, A.K.; Coutinho, C.G.; Pinheiro, K.M.; Aguiar, A.C.F. Management of a cohesive tropical soil to enhance rootability and increase the efficiency of nitrogen and potassium use. *Soil Use Manag.* **2012**, *28*, 370–377. [CrossRef]

24. Malhi, Y. The productivity, metabolism and carbon cycle of tropical forest vegetation. *J. Ecol.* **2012**, *100*, 65–75. [CrossRef]

25. Coelho, K.P.; Ribeiro, P.R.D.A.; Moura, E.G.D.; Aguiar, A.D.C.F.; Rodrigues, T.L.; Moreira, F.M.D.S. Symbiosis of rhizobia with Gliricidia sepium and Clitoria fairchildiana in an Oxisol in the pre-Amazon region of Maranhão State. *Acta Sci. Agron.* **2018**, *40*, e35248. [CrossRef]

26. Graham, P.H.; Vance, C.P. Nitrogen fixation in perspective: An overview of research and extension needs. *Field Crops Res.* **2000**, *65*, 93–106. [CrossRef]

27. Zhang, X.; Bol, R.; Rahn, C.; Xiao, G.; Meng, F.; Wu, W. Agricultural sustainable intensification improved nitrogen use efficiency and maintained high crop yield during 1980–2014 in Northern China. *Sci. Total Environ.* **2017**, *596–597*, 61–68. [CrossRef]

28. Bruijn, F.J. Biological nitrogen fixation. In *Principles of Plant-Microbe Interactions*; Lugtenberg, B., Ed.; Springer: Cham, Switzerland, 2015; pp. 215–224.

29. Vitousek, P.M.; Menge, D.N.; Reed, S.C.; Cleveland, C.C. Biological nitrogen fixation: Rates, patterns and ecological controls in terrestrial ecosystems. *Philos. Trans. R. Soc. B Biol. Sci.* **2013**, *368*, 2013–2019. [CrossRef]

30. Lima, A.S.; Nóbrega, R.S.A.; Barberi, A.; Silva, K.; Ferreira, D.F.; Moreira, F.M.S. Nitrogenfixing bacteria communities occurring in soils under different uses in the Western Amazon Region as indicated by nodulation of siratro (*Macroptilium atropurpureum*). *Plant Soil* **2009**, *319*, 127–145. [CrossRef]

31. Guimarães, A.A.; Jaramillo, P.M.D.; Nóbrega, R.S.A.; Florentino, L.A.; Silva, K.B.; Moreira, F.M.S. Genetic and symbiotic diversity of nitrogen-fixing bacteria isolated from agricultural soils in the Western Amazon by using cowpea as the trap plant. *Appl. Environ. Microbiol.* **2012**, *78*, 6726–6733. [CrossRef]

32. Santos, M.S.; Nogueira, M.A.; Hungria, M. Microbial inoculants: Reviewing the past, discussing the present and previewing an outstanding future for the use of beneficial bacteria in agriculture. *AMB Express* **2019**, *9*, 205. [CrossRef] [PubMed]

33. Biswas, B.; Scott, P.T.; Gresshoff, P.M. Tree legumes as feedstock for sustainable biofuel production: Opportunities and challenges. *J. Plant Physiol.* **2011**, *168*, 1877–1884. [CrossRef] [PubMed]

34. Reis, V.M.; Olivares, F.L. *Vias de Penetração e Infecção de Plantas por Bactérias*; EMBRAPA Agrobiologia: Seropédica, Brazil, 2006; pp. 1–34.

35. Suzaki, T.; Yoro, E.; Kawaguchi, M. Leguminous plants: Inventors of root nodules to accommodate symbiotic bacteria. In *International Review of Cell and Molecular Biology*; Jeon, K.W., Ed.; Academic Press: Cambridge, MA, USA, 2015; Volume 304, pp. 111–158.

36. Liu, C.W.; Murray, J.D. The role of flavonoids in nodulation host-range specificity: An update. *Plants* **2016**, *5*, 33. [CrossRef] [PubMed]

37. Trabelsi, D.; Mhamdi, R. Microbial inoculants and their impact on soil microbial communities: A review. *BioMed Res. Int.* **2013**. [CrossRef] [PubMed]

38. Bakhoum, N.; Ndoye, F.; Kane, A.; Assigbetse, K.; Fall, D.; Sylla, S.N.; Diouf, D. Impact of rhizobial inoculation on Acacia senegal (L.) Willd. growth in greenhouse and soil functioning in relation to seed provenance and soil origin. *World J. Microbiol. Biotechnol.* **2012**, *28*, 2567–2579. [CrossRef] [PubMed]

39. Thilakarathna, M.S.; Raizada, M.N. A meta-analysis of the effectiveness of diverse rhizobia inoculants on soybean traits under field conditions. *Soil Biol. Biochem.* **2017**, *105*, 177–196. [CrossRef]

40. Lindström, K.; Mousavi, S.A. Effectiveness of nitrogen fixation in rhizobia. *Microb Biotechnol.* **2019**. [CrossRef]

41. Moreira, F.M.S.; Siqueira, J.O. *Microbiologia e Bioquímica do Solo*, 2nd ed.; Univesidade Federal de Lavras: Lavras, Brazil, 2006; p. 729.

42. Masson-Boivin, C.; Giraud, E.; Perret, X.; Batut, J. Establishing nitrogen-fixing symbiosis with legumes: How many rhizobium recipes? *Trends Microbiol.* **2009**, *17*, 10. [CrossRef]

43. Perret, X.; Staehelin, C.; Broughton, W.J. Molecular basis of symbiotic promiscuity. *Microbiol. Mol. Biol. Rev.* **2000**, *64*, 180–201. [CrossRef]

44. Sachs, J.L.; Quides, K.W.; Wendlandt, C.E. Legumes versus rhizobia: A model for ongoing conflict in symbiosis. *New Phytol.* **2018**, *219*, 1199–1206. [CrossRef]

45. Taylor, B.N.; Simms, E.L.; Komats, K.J. More Than a Functional Group: Diversity within the Legume–Rhizobia Mutualism and Its Relationship with Ecosystem Function. *Diversity* **2020**, *12*, 50. [CrossRef]

46. Ramoneda, J.; Roux, J.L.; Frossard, E.; Frey, B.; Gamper, H.A. Different ecological processes drive the assembly of dominant and rare root-associated bacteria in a promiscuous legume. *bioRxiv* **2020**. [CrossRef]

47. Andrews, M.; Andrews, M.E. Specificity in Legume-Rhizobia Symbioses. *Int. J. Mol. Sci.* **2017**, *18*, 705. [CrossRef] [PubMed]

48. Ehinger, M.; Mohr, T.J.; Starcevich, J.B.; Sachs, J.L.; Porter, S.S.; Simms, E.L. Specialization-generalization trade-off in a *Bradyrhizobium* symbiosis with wild legume hosts. *BMC Ecol.* **2014**, *14*, 8. [CrossRef]

49. Martinez-Romero, E. Diversity of Rhizobium-Phaseolus vulgaris symbiosis: Overview and perspectives. *Plant Soil.* **2003**, *252*, 11–23. [CrossRef]

50. Jain, P.; Pundir, R.K. Recent Trends in Identification and Molecular Characterization of Rhizobia Species. In *Rhizobium Biology and Biotechnology*; Hansen, A.P., Choudhary, D.K., Agrawal, P.K., Varma, A., Eds.; Springer: Cham, Switzerland, 2017; Volume 50, pp. 136–164.

51. Wang, E.T.; Chen, W.F.; Tian, C.F.; Young, J.P.W.; Chen, W.X. *Ecology and Evolution of Rhizobia: Principles and Applications*; Springer: Cham, Switzerland, 2019; (eBook).

52. Suneja, P.; Duhan, J.S.; Bhutani, N.; Dudeja, S.S. Recent Biotechnological Approaches to Study Taxonomy of Legume Nodule Forming Rhizobia. In *Plant Biotechnology: Recent Advancements and Developments*; Gahlawat, S.K., Salar, R.K., Siwach, P., Duhan, J.S., Kumar, S., Kaur, P., Eds.; Springer: Cham, Switzerland, 2017; pp. 135–164.

53. Case, R.J.; Boucher, Y.; Dahllöf, I.; Holmström, C.; Doolittle, W.F.; Kjelleberg, S. Use of 16S rRNA and *rpoB* genes as molecular markers for microbial ecology studies. *Appl. Environ. Microbiol.* **2007**, *73*, 73278–73288. [CrossRef]

54. Salvina, L.H.O.S.; Rus, A.; Borozan, A.; Popescu, S. Preliminary studies regarding the development of a procedure for genetic identification of Rhizobium species. *J. Hortic. For.* **2018**, *2*, 94–99.

55. Hakim, S.; Mirza, B.S.; Imran, A.; Zaheer, A.; Yasmin, S.; Mubeen, F.; Mclean, J.E.; Mirza, M.S. Illumina sequencing of 16S rRNA tag shows disparity in rhizobial and nonrhizobial diversity associated with root nodules of mung bean (*Vigna radiate* L.) growing in different habitats in Pakistan. *Microbiol. Res.* **2020**, *231*, 126356. [CrossRef] [PubMed]

56. Azevedo, H.; Lopes, F.M.; Silla, P.R.; Hungria, M.A. database for the taxonomic and phylogenetic identification of the genus *Bradyrhizobium* using multilocus sequence analysis. *BMC Genom.* **2015**, *16*, 1–10. [CrossRef]

57. Hungria, M.; Menna, P.; Delamuta, J.R.M. *Bradyrhizobium*, the ancestor of all rhizobia: Phylogeny of housekeeping and nitrogen-fixation genes. In *Biological Nitrogen Fixation*; De BRUIJN, F.J., Ed.; Wiley Blackwell: Hoboken, NJ, USA, 2015; pp. 191–202.

58. Lambais, M.R.; Cury, J.D.C.; Maluche-Baretta, C.R.; Büll, R.D.C. Diversidade microbiana nos solos: Definindo novos paradigmas. *Tópicos em Ciência do Solo* **2005**, *4*, 43–84.

59. Jesus, E.C.; Moreira, F.M.S.; Florentino, L.A.; Rodrigues, M.I.D.; Oliveira, M.S. Diversidade de bactérias que nodulam siratro em três sistemas de uso da terra da Amazônia Ocidental. *Pesqui. Agropecu. Bras.* **2005**, *40*, 769–776. [CrossRef]

60. Vandamme, P.; Pot, B.; Gillis, M.; De Vos, P.; Kersters, K.; Swings, J. Polyphasic taxonomy, a consensus approach to bacterial systematics. *Microbiol. Rev.* **1996**, *60*, 407–438. [CrossRef] [PubMed]

61. Vincent, J.M.A. *Manual for the Practical Study of Root-Nodule Bacteria*; Blackwell Scientific: Oxford, UK, 1970; p. 164.

62. Moreira, F.M.S.; Pereira, E.G. Microsymbionts: Rhizobia. In *Standard Methods for Assessment of Soil Biodiversity and Land Use Practice*; SWIFT, M., BIGNELL, D., Eds.; International Centre for Research in Agroforestry: Bogor, Indonesia, 2001; pp. 19–24.

63. Thomas-Oates, J.; Bereszczak, J.; Edwards, E.; Gill, A.; Noreen, S.; Zhou, J.C.; Chen, M.Z.; Miao, L.H.; Xie, F.L.; Yang, J.K.; et al. A catalogue of molecular, physiological and symbiotic properties of soybean-nodulating rhizobial strains from different soybean cropping areas of China. *Syst. Appl. Microbiol.* **2003**, *26*, 453–465. [CrossRef] [PubMed]

64. Grönemeyer, J.L.; Bünger, W.; Reinhold-Hurek, B. *Bradyrhizobium namibiense* sp. nov., a symbiotic nitrogen-fixing bacterium from root nodules of Lablab purpureus, hyacinth bean, in Namibia. *Int. J. Syst. Evol. Microbiol.* **2017**, *67*, 4884–4891. [PubMed]

65. Safronova, V.I.; Sazanova, A.L.; Kuznetsova, I.G.; Belimov, A.A.; Andronov, E.E.; Chirak, E.R.; Popova, J.P.; Verkhozina, A.V.; Willems, A.; Tikhonovich, I.A. *Phyllobacterium zundukense* sp. nov., a novel species of rhizobia isolated from root nodules of the legume species *Oxytropis triphylla* (Pall.) Pers. *Int. J. Syst. Evol. Microbiol.* **2018**, *68*, 1644–1651. [CrossRef] [PubMed]

66. Frank, B. Über die Pilzsymbiose der Leguminosen. *Berichte Dtsch. Bot. Ges.* **1889**, *7*, 332–346.

67. Casida Junior, L.E. *Ensifer adhaerens* gen. nov., sp. nov.: A bacterial predator of bacteria in soil. *Int. J. Syst. Bacteriol.* **1982**, *32*, 339–345.

68. Chen, W.X.; Yan, G.H.; Li, J.L. Numerical taxonomic study of fast-growing soybean rhizobia and a proposal that *Rhizobium fredii* be assigned to *Sinorhizobium* gen. nov. *Int. J. Syst. Bacteriol.* **1988**, *38*, 392–397. [CrossRef]

69. De Lajudie, P.; Laurent-Fulele, E.; Willems, A.; Torek, U.; Coopman, R.; Collins, M.D.; Kersters, K.; Dreyfus, B.; Gillis, M. *Allorhizobium undicola* gen. nov., sp. nov., nitrogen-fixing bacteria that efficiently nodulate *Neptunia natans* in Senegal. *Int. J. Syst. Bacteriol.* **1998**, *48*, 1277–1290. [CrossRef]

70. Ren, D.W.; Wang, E.T.; Chen, W.F.; Sui, X.H.; Zhang, X.X.; Liu, H.C.; Chen, W.X. *Rhizobium herbae* sp. nov. and *Rhizobium giardinii*-related bacteria, minor microsymbionts of various wild legumes in China. *Int. J. Syst. Evol. Microbiol.* **2011**, *61*, 1912–1920. [CrossRef]

71. An, D.S.; Im, W.T.; Yang, H.C.; Lee, S.T. *Shinella granuli* gen. nov., sp. nov., and proposal of the reclassification of Zoogloea ramigera ATCC 19623 as *Shinella zoogloeoides* sp. nov. *Int. J. Syst. Evol. Microbiol* **2006**, *56*, 443–448X. [CrossRef] [PubMed]

72. Lin, D.X. *Shinella kummerowiae* sp. nov., a symbiotic bacterium isolated from root nodules of the herbal legume *Kummerowia stipulacea*. *Int. J. Syst. Evol. Microbiol.* **2008**, *58*, 1409–1413. [CrossRef] [PubMed]

73. Mousavi, S.A.; Österman, J.; Wahlberg, N.; Nesme, X.; Lavire, C.; VialC, L.; Paulin, L.; De Lajudie, P.; Lindström, K. Phylogeny of the *Rhizobium–Allorhizobium–Agrobacterium* cladesupports the delineation of *Neorhizobium* gen. nov. *Syst. Appl. Microbiol.* **2014**, *37*, 208–215. [CrossRef] [PubMed]

74. Dreyfus, B.; Garcia, J.L.; Gillis, M. Characterization of *Azorhizobium caulinodans* gen. nov., sp. nov., a stem-nodulating nitrogen-fixing bacterium isolated from *Sesbania rostrata*. *Int. J. Syst. Bacteriol.* **1988**, *38*, 89–98. [CrossRef]

75. Lang, E.; Schumann, P.; Adler, S.; Spröer, C.; Sahin, N. *Azorhizobium oxalatiphilum* sp. nov., and emended description of the genus *Azorhizobium*. *Int. J. Syst. Evol. Microbiol.* **2013**, *63*, 1505–1511. [CrossRef] [PubMed]

76. Moreira, F.M.S.; Cruz, L.; De Faria, S.M.; Marsh, T.; Martínez-Romero, E.; de Oliveira Pedrosa, F.; Pitard, R.M.; Young, J.P.W. *Azorhizobium doebereinerae* sp. nov. Microsymbiont of *Sesbania virgata* (Caz.) Pers. *Syst. Appl. Microbiol.* **2006**, *29*, 197–206.

77. Nakagawa, Y.; Sakane, T.; Yokota, A. Transfer of "*Pseudomonas riboflavina*" (Foster 1944), a gram-negative, motile rod with long-chain 3-hydroxy fatty acids, to Devosia riboflavina gen. nov., sp. nov., nom. rev. *Int. J. Syst. Bacteriol* **1996**, *46*, 16–22. [CrossRef]

78. Rivas, R.; Willems, A.; Subba-Rao, N.S.; Mateos, P.F.; Dazzo, F.B.; Kroppenstedt, R.M.; Martínez-Molina, E.; Gillis, M.; Velázquez, E. Description of *Devosia neptuniae* sp. nov. that nodulates and fixes nitrogen in symbiosis with *Neptunia natans*, an aquatic legume from India. *Syst. Appl. Microbiol.* **2003**, *26*, 47–53. [CrossRef]

79. Jordan, D.C. Transfer of *Rhizobium japonicum* Buchanan 1980 to *Bradyrhizobium* gen. nov., a genus of slow-growing, root nodule bacteria from leguminous plants. *Int. J. Syst. Bacteriol.* **1982**, *32*, 136–139.

80. Buchanan, R.E. Approved lists of bacterial names. *Int. J. Syst. Bacteriol.* **1980**, *30*, 225–420.

81. Kirchner, O. Die Wurzelknöllchen der Sojabohne. *Beiträge zur Biologie der Pflanzen* **1896**, *7*, 213–224.

82. Van Berkum, P.; Eardly, B.D. The aquatic budding bacterium *Blastobacter denitrificans* is a nitrogen-fixing symbiont of *Aeschynomene indica*. *Appl. Environ. Microbiol.* **2002**, *68*, 1132–1136. [CrossRef] [PubMed]

83. Jarvis, B.D.W.; Van Berkum, P.; Chen, W.X.; Nour, S.M.; Fernandez, M.P.; Cleyet-Marel, J.C.; Gillis, M. Transfer of *Rhizobium loti, Rhizobium huakuii, Rhizobium ciceri, Rhizobium mediterraneum*, and *Rhizobium tianshanense* to *Mesorhizobium* gen. nov. *Int. J. Syst. Bacteriol.* **1997**, *47*, 895–898. [CrossRef]

84. Jarvis, B.D.W.; Pankhurst, C.E.; Patel, J.J. *Rhizobium loti*, a new species of legume root nodule bacteria. *Int. J. Syst. Bacteriol. Read.* **1982**, *32*, 378–380. [CrossRef]

85. Urakami, T.; Araki, H.; Oyanagi, H.; Suzuki, K.I.; Komagata, K. Transfer of *Pseudomonas aminovorans* (den Dooren de Jong 1926) to *Aminobacter* gen. nov. as *Aminobacter aminovorans* comb. nov. and description of *Aminobacter aganoensis* sp. nov. and *Aminobacter niigataensis* sp. nov. *Int. J. Syst. Bacteriol.* **1992**, *42*, 84–92.

86. Maynaud, G.; Willems, A.; Soussou, S.; Vidal, C.; Mauré, L.; Moulin, L.; Cleyet-Marel, J.; Brunel, B. Molecular and phenotypic characterization of strains nodulating *Anthyllis vulneraria* in mine tailings, and proposal of *Aminobacter anthyllidis* sp. nov., the first definition of *Aminobacter* as legume-nodulating bacteria. *Syst. Appl. Microbiol.* **2012**, *35*, 65–72. [CrossRef] [PubMed]

87. Zimmermann, A. Uber Bakterienknoten in den Bla¨ttern einiger Rubiaceen. *Jahrb. Wiss. Bot.* **1902**, *37*, 1–11.

88. Knösel, D.H. Genus *Phyllobacterium*. In *Bergey's Manual of Systematic Bacteriology*; Krieg, N.R., Holt, J.G., Eds.; The Williams & Wilkins Co.: Baltimore, MD, USA, 1984; Volume 1, pp. 254–256.

89. Patt, T.E.; Cole, G.C.; Hanson, R.S. *Methylobacterium*, a new genus of facultatively methylotrophic bacteria. *Int. J. Syst. Bacteriol.* **1976**, *26*, 226–229. [CrossRef]

90. Sy, A.; Giraud, E.; Jourand, P.; Garcia, N.; Willems, A.; De Lajudie, P.; Prin, Y.; Neyra, M.; Gillis, M.; Boivin-Masson, C.; et al. Methylotrophic *Methylobacterium* bacteria nodulate and fix nitrogen in symbiosis with legumes. *J. Bacteriol.* **2001**, *183*, 214–220. [CrossRef] [PubMed]

91. Kanso, S.; Patel, B.K. *Microvirga subterranea* gen. nov., sp. nov., a moderate thermophile from a deep subsurface Australian thermal aquifer. *Int. J. Syst. Evol. Microbiol.* **2003**, *53*, 401–406. [CrossRef]

92. Radl, V.; Simões-Araújo, J.L.; Leite, J.; Passos, S.R.; Martins, L.M.; Xavier, G.R.; Rumjanek, N.G.; Baldani, J.I.; Zilli, J.E. Microvirga vignae sp. nov., a root nodule symbiotic bacterium isolated from cowpea grown in semi-arid Brazil". *Int. J. Syst. Evol. Microbiol.* **2014**, *64*, 725–730. [CrossRef] [PubMed]

93. Holmes, B.; Popoff, M.; Kiredjian, M.; Kersters, K. *Ochrobactrum anthropi* gen. nov., sp. nov. from human clinical specimens and previously known as group Vd. *Int. J. Syst. Bacteriol.* **1988**, *38*, 406–416.

94. Ngom, A.; Nakagawa, Y.; Sawada, H.; Tsukahara, J.; Wakabayashi, S.; Uchiumi, T.; Nuntagij, A.; Kotepong, S.; Suzuki, A.; Higashi, S.; et al. A novel symbiotic nitrogen-fixing member of the *Ochrobactrum* clade isolated from root nodules of *Acacia mangium*. *J. Gen. Appl. Microbiol.* **2004**, *50*, 17–27. [CrossRef] [PubMed]

95. Trujillo, M.E.; Willems, A.; Abril, A.; Planchuelo, A.M.; Rivas, R.; Ludena, D.; Mateos, P.F.; Martinez-Molina, E.; Velázquez, E. Nodulation of *Lupinus albus* by Strains of *Ochrobactrum lupini* sp. nov. *Appl. Environ. Microbiol.* **2005**, *71*, 1318–1327. [CrossRef] [PubMed]

96. Zurdo-Pineiro, J.L.; Rivas, R.; Trujillo, M.E.; Vizcaino, N.; Carrasco, J.A.; Chamber, M.; Palomares, A.; Mateos, P.F.; Martinez-Molina, E.; Velazquez, E. *Ochrobactrum cytisi* sp. nov., isolated from nodules of *Cytisus scoparius* in Spain. *Int. J. Syst. Evol. Microbiol.* **2007**, *57*, 784–788.

97. Yabuuchi, E.; Kosako, Y.; Oyaizu, H.; Yano, I.; Hotta, H.; Hashimoto, Y.; Ezaki, T.; Arakawa, M. Proposal of *Burkholderia* gen. nov. and transfer of seven species of the genus *Pseudomonas* homology group II to the new genus, with the type species *Burkholderia cepacia* (Palleroni and Holmes 1981) comb. nov. *Microbiol. Immunol.* **1992**, *36*, 1251–1275. [CrossRef]

98. Vandamme, P.; Goris, J.; Chen, W.M.; De Vos, P.; Willems, A.A. *Burkholderia tuberum* sp. nov. and *Burkholderia phymatum* sp. nov., nodulate the roots of tropical legumes. *Syst. Appl. Microbiol.* **2002**, *25*, 507–512.

99. Sawana, A.; Adeolu, M.; Gupta, R.S. Molecular signatures and phylogenomic analysis of the genus Burkholderia: Proposal for division of this genus into the emended genus Burkholderia containing pathogenic organisms and a new genus Paraburkholderia gen.nov.harboring environmental species. *Genetics* **2014**, *5*, 429. [CrossRef]

100. Makkar, N.S.; Casida Junior, L.E. *Cupriavidus necator* gen. nov., sp. nov.; a nonobligate bacterial predator of bacteria in soil. *Int. J. Syst. Bacteriol.* **1987**, *37*, 323–326. [CrossRef]

101. Barrett, C.F.; Parker, M.A. Coexistence of *Burkholderia*, *Cupriavidus*, and *Rhizobium sp.* nodule Bacteria on two *Mimosa spp.* in Costa Rica. *Appl. Environ. Microbiol.* **2006**, *72*, 1198–1206. [CrossRef]

102. Silva, K.; Florentinho, L.A.; Silva, K.B.; Brandt, E.; Vandamme, P.; Moreira, F.M.S. *Cupriavidus necator* isolates are able to fix nitrogen in symbiosis with different legume species. *Syst. Appl. Microbiol.* **2012**, *35*, 175–182. [PubMed]

103. Yabuuchi, E.; Kosako, Y.; Yano, I.; Hotta, H.; Nishiuchi, Y. Transfer of two *Burkholderia* and an *Alcaligenes* species to *Ralstonia* gen. nov.: Proposal of *Ralstonia pickettii* (Ralston, Palleroni and Doudoroff 1973) comb. nov., *Ralstonia solanacearum* (Smith 1896) comb. nov. and *Ralstonia eutropha* (Davis 1969) comb. nov. *Microbiol. Immunol.* **1995**, *39*, 897–904. [CrossRef] [PubMed]

104. Chen, W.M.; Laevens, S.; Lee, T.M.; Coenye, T.; De Vos, P.; Mergeay, M.; Vandamme, P. *Ralstonia taiwanensis* sp. nov., isolated from root nodules of *Mimosa* species and sputum of a cystic fibrosis patient. *Int. J. Syst. Evol. Microbiol.* **2001**, *51*, 1729–1735. [CrossRef] [PubMed]

105. Bala, A.; Giller, K.E. Relationships between rhizobial diversity and host legume nodulation and nitrogen fixation in tropical ecosystems. *Nutr. Cycling Agroecosyst.* **2006**, *76*, 319–330. [CrossRef]

106. Bala, A.; Murphy, P.J.; Osunde, A.O.; Giller, K.E. Nodulation of tree legumes and the ecology of their native rhizobial populations in tropical soils. *Appl. Soil Ecol.* **2003**, *22*, 211–223. [CrossRef]

107. Nascimento, J.T.; Silva, I.F. Avaliação quantitativa e qualitativa da fitomassa de leguminosas para uso como cobertura de solo. *Cienc. Rural* **2004**, *34*, 947–949. [CrossRef]

108. Costa, J.N.M.N.; Durigan, G. *Leucaena leucocephala* (Lam.) de Wit (Fabaceae): Invasive or ruderal? *Rev. Arvore* **2010**, *34*, 825–833. [CrossRef]

109. Bala, A.; Giller, K.E. Symbiotic specificity of tropical tree rhizobia for host legumes. *New Phytol.* **2001**, *149*, 495–507. [CrossRef]

110. Bala, A.; Murphy, P.; Giller, K.E. Distribution and diversity of rhizobia nodulating agroforestry legumes in soils from three continents in the tropics. *Mol. Ecol.* **2003**, *12*, 917–929. [CrossRef]

111. Xu, K.W.; Penttinen, P.; Chen, Y.X.; Chen, Q.; Zhang, X. Symbiotic efficiency and phylogeny of the rhizobia isolated from Leucaena leucocephala in arid–hot river valley area in Panxi, Sichuan, China. *Appl. Microbiol. Biotechnol.* **2013**, *97*, 783–793. [CrossRef]

112. Wang, E.T.; Kan, F.L.; Tan, Z.Y.; Toledo, I.; Chen, W.X.; Martínez-Romero, E. Diverse *Mesorhizobium plurifarium* populations native to Mexican soils. *Arch. Microbiol.* **2003**, *180*, 444–454. [CrossRef] [PubMed]

113. Florentino, L.A.; Guimarães, A.P.; Rufini, M.; Silva, K.; Moreira, F.M.S. *Sesbania virgata* stimulates the occurrence of its microsymbiont in soils but does not inhibit microsymbionts of other species. *Sci. Agric.* **2009**, *66*, 667–676. [CrossRef]

114. Moreira, F.M.S.; Gillis, M.; Pot, B.; Kersters, K.; Franco, A.A. Characterization of rhizobia isolated from different divergence groups of tropical Leguminosae by comparative polyacrylamide gel electrophoresis of their total proteins. *Syst. Appl. Microbiol.* **1993**, *16*, 135–146. [CrossRef]

115. Wang, F.Q.; Wang, E.T.; Zhang, Y.F.; Chen, W.X. Characterization of rhizobia isolated from *Albizia* spp. in comparison with microsymbionts of *Acacia* spp. and *Leucaena leucocephala* grown in China. *Syst. Appl. Microbiol.* **2006**, *29*, 502–517. [CrossRef]

116. Clapp, J.P.; Mansur, I.; Dodd, J.C.; Jeffries, P. Ribotyping of rhizobia nodulating *Acacia mangium* and *Paraserianthes falcataria* from differentes geografical areas in Indonesia using PCR-RFLP-SSCP (PRS) and sequencing. *Environ. Microbiol.* **2001**, *3*, 273–280. [CrossRef] [PubMed]

117. Le Roux, C.; Tentchev, D.; Prin, Y.; Goh, D.; Japarudin, Y.; Perrineau, M.M.; Galiana, A. *Bradyrhizobia* nodulating the *Acacia mangium×* *A. auriculiformis* interspecific hybrid are specific and differ from those associated with both parental species. *Appl. Environ. Microbiol.* **2009**, *75*, 7752–7759. [CrossRef] [PubMed]

118. Perrineau, M.M.; Le Roux, C.; De Faria, S.M.; De Carvalho Balieiro, F.; Galiana, A.; Prin, Y.; Béna, G. Genetic diversity of symbiotic *Bradyrhizobium elkanii* populations recovered from inoculated and non-inoculated *Acacia mangium* field trials in Brazil. *Syst. Appl. Microbiol.* **2011**, *34*, 376–384. [CrossRef]

119. Perrineau, M.M.; Le Roux, C.; Galiana, A.; Faye, A.; Duponnois, R.; Goh, D.; Béna, G. Differing courses of genetic evolution of *Bradyrhizobium* inoculants as revealed by long-term molecular tracing in *Acacia mangium* plantations. *Appl. Environ. Microbiol.* **2014**, *80*, 5709–5716. [CrossRef]

120. Moreira, F.M.S.; Haukka, K.; Young, J.P.W. Biodiversity of rhizobia isolated from a wide range of forest legumes in Brazil. *Mol. Ecol.* **1998**, *7*, 889–895. [CrossRef]

121. Bray, R.A. Diversity within tropical tree and shrub legumes. In *Forage Tree Legumes in Tropical Agriculture*; Mathison, G.W., Ed.; CAB International: Wallingford, UK, 1994; pp. 101–106.

122. Batish, D.R.; Kohli, R.K.; Jose, S.; Singh, H.P. (Eds.) *Ecological Basis of Agroforestry*; CRC Press: Boca Raton, FL, USA, 2007.

123. Acosta-Durán, C.; Martínez-Romero, E. Diversity of rhizobia from nodules of the leguminous tree *Gliricidia sepium*, a natural host of *Rhizobium tropici*. *Arch. Microbiol.* **2002**, *178*, 161–164. [CrossRef]

124. Florentino, L.A.; Rezende, A.V.; Mesquita, A.C.; Lima, A.R.; Marques, D.J.; Miranda, J.M. Diversidade e potencial de utilização dos rizóbios isolados de nódulos de *Gliricidia sepium*. *Revista de Ciências Agrárias* **2014**, *37*, 320–338.

125. Degefu, T.; Wolde-Meskel, E.; Frostegård, Å. Phylogenetic diversity of *Rhizobium* strains nodulating diverse legume species growing in Ethiopia. *Syst. Appl. Microbiol.* **2013**, *36*, 272–280. [CrossRef] [PubMed]
126. Lorenzi, H. *Árvores brasileiras: Manual de Identificação e Cultivo de Plantas Arbóreas Nativas do Brasil*; Plantarum: Nova Odessa, Brazil, 1992.
127. Leite, A.A.L.; Ferraz Junior, A.S.L.; Moura, E.G.; Aguiar, A.C.F. Comportamento de dois genótipos de milho cultivados em sistema de aléias pré-estabelecidos com diferentes leguminosas arbóreas. *Bragantia* **2008**, *67*, 875–882. [CrossRef]

MDPI

St. Alban-Anlage 66

4052 Basel

Switzerland

Tel. +41 61 683 77 34

Fax +41 61 302 89 18

www.mdpi.com

Diversity Editorial Office

E-mail: diversity@mdpi.com

www.mdpi.com/journal/diversity